12-26

Biological Activities of Polymers

Biological Activities of Polymers

Charles E. Carraher, Jr., EDITOR
Wright State University

Charles G. Gebelein, EDITOR
Youngstown State University

Based on a symposium sponsored by the Division of Organic Coatings and Plastics Chemistry at the 181st Meeting of the American Chemical Society, Atlanta, Georgia, March 30–31, 1981.

ACS SYMPOSIUM SERIES **186**

AMERICAN CHEMICAL SOCIETY
WASHINGTON, D. C. 1982

Library of Congress CIP Data
Biological activities of polymers.
(ACS symposium series, ISSN 0097–6156; 186)

Includes bibliography and index.

1. Macromolecules—Physiological effect. 2. Macromolecules—Therapeutic use. 3. Polymers and polymerization.
I. Carraher, Charles E. II. Gebelein, Charles G. III. Series.

QP801.P64B55   615'.3   82–3988
ISBN 0-8412-0719-4 AACR2 ACSMC8 186 1–293
                                              1982

Copyright © 1982

American Chemical Society

All Rights Reserved. The appearance of the code at the bottom of the first page of each article in this volume indicates the copyright owner's consent that reprographic copies of the article may be made for personal or internal use or for the personal or internal use of specific clients. This consent is given on the condition, however, that the copier pay the stated per copy fee through the Copyright Clearance Center, Inc. for copying beyond that permitted by Sections 107 or 108 of the U.S. Copyright Law. This consent does not extend to copying or transmission by any means—graphic or electronic—for any other purpose, such as for general distribution, for advertising or promotional purposes, for creating new collective work, for resale, or for information storage and retrieval systems.

The citation of trade names and/or names of manufacturers in this publication is not to be construed as an endorsement or as approval by ACS of the commercial products or services referenced herein; nor should the mere reference herein to any drawing, specification, chemical process, or other data be regarded as a license or as a conveyance of any right or permission, to the holder, reader, or any other person or corporation, to manufacture, reproduce, use, or sell any patented invention or copyrighted work that may in any way be related thereto.

PRINTED IN THE UNITED STATES OF AMERICA

ACS Symposium Series

M. Joan Comstock, *Series Editor*

Advisory Board

David L. Allara	Marvin Margoshes
Robert Baker	Robert Ory
Donald D. Dollberg	Leon Petrakis
Robert E. Feeney	Theodore Provder
Brian M. Harney	Charles N. Satterfield
W. Jeffrey Howe	Dennis Schuetzle
James D. Idol, Jr.	Davis L. Temple, Jr.
Herbert D. Kaesz	Gunter Zweig

FOREWORD

The ACS SYMPOSIUM SERIES was founded in 1974 to provide a medium for publishing symposia quickly in book form. The format of the Series parallels that of the continuing ADVANCES IN CHEMISTRY SERIES except that in order to save time the papers are not typeset but are reproduced as they are submitted by the authors in camera-ready form. Papers are reviewed under the supervision of the Editors with the assistance of the Series Advisory Board and are selected to maintain the integrity of the symposia; however, verbatim reproductions of previously published papers are not accepted. Both reviews and reports of research are acceptable since symposia may embrace both types of presentation.

CONTENTS

Preface ... ix

1. Perspectives in Bioactive Polymers ... 1
   C. G. Gebelein and C. E. Carraher, Jr.

NONMEDICAL APPLICATIONS

2. Biological Activities of Metal-Containing Polymers ... 13
   C. E. Carraher, Jr., D. J. Giron, D. R. Cerutis, W. R. Burt,
   R. S. Venkatachalam, T. J. Gehrke, S. Tsuji, and H. S. Blaxall

3. Biocidal Activity of Organotin Polymers in Wood ... 27
   D. M. Andersen, J. A. Mendoza, B. K. Garg,
   and R. V. Subramanian

4. Polymer-Bound Fungicides for Paints ... 35
   C. U. Pittman, Jr., K. R. Lawyer, and K. S. Ramachandran

5. Poly(thiosemicarbazide) Copper(II) Complexes as Potential
   Algicides and Molluscicides ... 55
   L. G. Donaruma, S. Kitoh, J. V. Depinto, J. K. Edzwald,
   and M. J. Maslyn

6. Potential Polymeric Herbicides Derived from Poly(vinyl alcohol):
   Modification of Polymers ... 75
   C. G. Gebelein

DRUG RELATED ACTIVITY: MEDICAL APPLICATIONS

7. In Vivo Studies on Drug–Polymer Sustained-Release Systems ... 85
   J. M. Anderson

8. Polymeric Delivery Systems for Macromolecules: Approaches for
   Studying In Vivo Release Kinetics and Designing Constant Rate
   Systems ... 95
   R. Langer, D. S. T. Hsieh, and L. Brown

9. Design of Polymeric Iron Chelators for Treating Iron Overload in
   Cooley's Anemia ... 107
   A. Winston, J. Rosthauser, D. Fair, J. Bapasola,
   and W. Lerdthusnee

10. Structural and Functional Interrelationships of Anterior Pituitary
    Hormones ... 119
    J. Ramachandran

11. Covalent Binding of Trypsin to Hydrogels ... 133
    W. H. Daly and F. Shih

12. **Chitin–Protein Complexes: Ordered Biopolymer Composites** **149**
    J. Blackwell, L. T. Germinario, and M. A. Weih

13. **Interaction of Synthetic Polymers with Cell Membranes and Model Membrane Systems: Pyran Copolymer** **163**
    L. K. Marwaha and D. A. Tirrell

14. **Thermodynamic Characterization of Proflavine Binding to DNA** ... **177**
    Y. Baba, C. L. Beatty, and A. Kagemoto

    ANTICANCER APPLICATIONS

15. **Polymers for Potential Cancer Therapy: A Brief Review** **193**
    C. G. Gebelein

16. **The Antitumor and Antiviral Effects of Polycarboxylic Acid Polymers** **205**
    R. M. Ottenbrite

17. **Polymeric Derivatives Based on cis-Diamminedichloroplatinum(II) as Antineoplastic Agents** **221**
    C. E. Carraher, Jr., W. J. Scott, I. Lopez, D. J. Giron, D. R. Cerutis, and T. Manek

18. **Controlled Release of Anticancer Agents That Are Complexes of cis-Diamminedichloroplatinum(II) and α-Hydroxyquinones** **233**
    S. Yolles, R. M. Roat, M. F. Sartori, and C. L. Washburne

19. **^{13}C NMR Studies of the Structure of the Divinyl Ether–Maleic Anhydride Cyclic Alternating Copolymer: A Biologically Active Agent** **243**
    W. J. Freeman and D. S. Breslow

20. **Interaction of Methotrexate (Polylysine) with Rat Liver Tumor Cells** **255**
    J. M. Whiteley and J. H. Galivan

21. **Evidence of Cooperativity and Allosterism in the Binding of Various Antibiotics and Carcinogens to DNA** **269**
    L. S. Rosenberg, M. S. Balakrishnan, D. E. Graves, K. R. Lee, S. A. Winkle, and T. R. Krugh

Index **285**

PREFACE

Biologically active polymers serve as the basis of life and have existed since the beginning of the first forms of life in an aqueous medium. The major macromolecular entities of all living creatures, both animal and plant, are biologically active polymers: proteins, nucleic acids, and polysaccharides. The recognition that some synthetic polymers also exhibit biological activity is of a more recent vintage. The term "biologically active polymers" or "bioactive polymers" encompasses synthetic and natural polymers (native or modified) capable of eliciting a physiological response when applied to or introduced into a living system. The utilization of the biological activities of these natural and/or synthetic polymers is a major and growing area of scientific endeavor (molecular engineering, polymeric drugs, etc.) and is the basis for this book.

The chapters of this volume were chosen to illustrate the broad spectrum of research currently being undertaken related to the biological activities of macromolecules. The initial chapter introduces the basic framework relating the biological activity of polymers to the test conditions, etc., and emphasizes the need to design tests for each specific purpose (i.e., match the tests with the potential end use of the bioactive polymer). The remainder of the book is divided into three sections, each focusing on the current state of the art in particular areas of biological activity. The first section considers a broad range of nonmedical applications, including treatment of wood, control of uranyl ion toxicity, and control of schistosomiasis-carrying snails. Also discussed are polymers acting as herbicides, pesticides, mildew- and rot-resisting agents, mollusicides, algicides, antifungal agents, and antibacterial agents.

The final two sections focus on some medically related applications of bioactive polymers, including targeting and controlled release of drugs. Types of polymer supports and structure–activity relationships are also well represented. The section on drug-related activity also contains chapters on polymer activity as applied to the treatment of specific diseases, and on the emergence of thermal analysis techniques to describe biological activities of polymers more rapidly. The last section is based on applications of biologically active polymers in the broad area of cancer research. Antitumor and antiviral activities are featured along with discussions involving the advantages of polymeric anticancer drugs over the low molecular weight analogous drugs.

We thank the authors for preparing the various manuscripts in this book and also, in some cases, for acting as reviewers. We also thank the Division of Organic Coatings and Plastics Chemistry for its help and encouragement in carrying out this project. In addition, we are grateful to the following who assisted in manuscript preparation, review, and general criticism: Waris Baig, Charles Carraher III, William Feld, Subrata Ghosh, Robert Hartsough, James Kane, David Karl, Robert Patsiga, and John Sheats. Finally, we thank our families for their support.

CHARLES E. CARRAHER, JR.
Wright State University
Department of Chemistry
Dayton, Ohio 45435

CHARLES G. GEBELEIN
Youngstown State University
Department of Chemistry
Youngstown, Ohio 44555

November 6, 1981.

Perspectives in Bioactive Polymers

CHARLES G. GEBELEIN
Youngstown State University, Department of Chemistry, Youngstown, OH 44555

CHARLES E. CARRAHER, JR.
Wright State University, Department of Chemistry, Dayton, OH 45435

>Bioactive polymers are macromolecules that can exhibit biological activity in use areas such as medication, weed control, insect control, etc. Natural bioactive polymers are essential to life and include the proteins, nucleic acids and polysaccharides. Synthetic bioactive polymers are a more recent development but hundreds of possible examples have been reported with potential biological activity. In this brief, introductory review, the history, philosophy, mode of activity and the advantages of bioactive polymers are discussed emphasizing synthetic polymers.

Biologically active polymers are of fundamental importance in the life process itself. We need only note that molecules such as the proteins, nucleic acids, enzymes and many complex sugars are macromolecular in nature to realize how dependent life is upon polymer molecules. In addition, many of the polypeptide hormones are derived from a larger polypeptide or are synthesized by means of other macromolecules. Nearly all our foods and most of our bodies are composed of polymers and most of these are bioactive. We can note with certainty that macromolecules are essential to life as we know it. Beyond this, the interaction of macromolecules plays an important role in many areas of life. Examples of this could include such divergent effects as the pain-killing activity of the enkephalins or the toxic nature of snake venom which is a polypeptide in most cases. In subsequent chapters of this book, other authors will consider a variety of polymeric systems with various types of actual or potential activity with humans, animals and/or plants. In this chapter, we briefly explore the overall realm of biologically active polymers.

0097-6156/82/0186-0001$5.00/0
© 1982 American Chemical Society

History

A complete, detailed history of biologically active polymers is beyond the present scope, but we can note that such a complete history would be very difficult to unravel since the threads of this story are entwined not only in the area of polymer science but also in fields such as biochemistry, pharmacology, molecular biology and medicine. We will restrict this brief survey to synthetic bioactive polymers. Even here, however, it is difficult to draw forth an accurate picture since the early chemical literature abounds with examples of monomers and polymers that could possibly exhibit biological activity. Some of this literature has been reviewed, at least in part, elsewhere (1,2). Some early examples of potentially biologically active polymers include 2-vinylthiophene in 1941 (3), 2-vinylfuran in 1946 (4), 2-vinylpyrrole in 1954 (5), and N-vinylpyrrolidone in 1952 (6). Since its synthesis, N-vinylpyrrolidone has been widely used as a bood plasma extender (although this use has been discontinued in the USA) and the aqueous iodine: PVP complex has been marketed as an antiseptic for many years.

The work began by Overberger in the 1950's was aimed at developing biologically active, synthetic polymers and involved the homopolymers and copolymers of the vinyl derivatives of pyrimidine and/or triazine compounds (7,8). Although the polymers had high molecular weights (up to 100,000), no biological activity was observed. In more recent work, Overberger and others have observed significant enzyme-like activity for some synthetic polymers and copolymers of vinyl imidazole derivaties (9,10). This pseudo-enzyme polymer area has been reviewed recently (11,12) and could prove to be of future importance in chemotherapy or in chemical manufacturing.

The major thrust in chemotherapeutic and other biologically active synthetic polymers began in the 1960's. While much of this work involved placing a known bioactive agent on or in a polymer chain, some biologically active polymers were developed which had no active, low molecular weight analogs. Hundreds of examples of such systems have been published by several dozen different investigating groups. We shall only examine a few specific examples and cite only some selected references. (Additional, related references can be found in these papers.)

Donaruma and coworkers have studied copolymers of various sulfa drugs with formaldehyde or dimethylolurea. These copolymers often exhibited bacteriostatic activity and in some cases this activity level was higher than with the corresponding, monomeric, sulfa drug. The polymers often showed lower toxicity levels and, in general, the dimethylolurea derivatives showed higher bacteriostatic activity than did the formaldehyde derivatives. In earlier work, Donaruma and coworkers also prepared some antimalarial, polymeric sulfomes and some bioactive tropone derivatives (1,13).

Ringsdorf and coworkers have prepared many monomers, polymers and copolymers that are potentially chemotherapeutic (14,15,16). These included derivatives of cyclophosphamide which could be a potential anticancer compound (15). In addition to homopolymers, they prepared water soluble copolymers with N-vinylpyrrolidone, vinyl amine oxides and other monomers. Other work by this group has included polymers containing steroids, sulfa drugs, chlorambucil and adamantylamine units. In his review article, Ringsdorf discusses the use of some advanced techniques such as a homing device (or directing units), solubilizing units and various spacer units along with the pharmacologically active unit. The purpose of the spacer unit is to move the bioactive unit out from the polymer backbone and to make it more readily available for chemotherapeutic activity. In some cases, the use of spacer units has greatly enhanced the activity of the polymeric drug (14).

Carraher, Pittman, McCormick, Subramanian, Sheats and others have synthesized a wide variety of metal and phosphorus containing polymers which exhibit biological activities (17-22). Tin polyamines, polyesters and polyethers all exhibit broad spectrum antibacterial and antifungal activities. Tin containing polysaccharides derived from a number of sources also exhibit good antifungal and antibacterial activities. Extensive studies have shown that tin containing vinyl polymers exhibit sustained antifouling properties and such materials are currently utilized in the control of barnacles on ship bottoms and shore installations.

Takemoto and coworkers have prepared a wide variety of monomeric derivatives of a number of nucleic acid bases in order to prepare simple analogs of RNA and DNA (23,24). Some of these synthetic polymers do show base-base interactions that are similar to those observed in natural DNA and RNA materials. Pitha has observed anticancer and antiviral activity for some vinyl derivative polymers of uracil or adenine (25).

Gebelein and coworkers have prepared monomers and polymers based on the antineoplastic agents 5-fluorouracil and 6-mercaptopurine (26). These polymers would be potentially useful in treating various types of cancer or leukemia and some derivatives have shown biological activity. Previous work by this group has included the synthesis of polyisoprene derivatives containing sulfa drug, carbamate, urea, oxazolidone and oxazole units (27,28).

Probably the most striking example of a biologically active polymer that does not have an active, low molecular weight analog is the 1:2 divinyl ether: maleic anhydride cyclic, alternating copolymer which is called 'pyran polymer' or DIVEMA. This polymer was first prepared by Butler in 1960 (29) and has been shown to exhibit a remarkable range of biological activity including antitumoral, antiviral, antibacterial, and antifungal activities (30). Another system which exhibits biological action as a polymer but not as a monomer are the polymers of 2-vinylpyridine-1-oxide which show antisilicosis activity (3). Several groups are attempting to develop synthetic polymers that would be capable of reversible oxygen binding and these polymers would lead to the development of artificial blood systmes (32,33). In addition to these bioactive polymer systems, much work has been done on the controlled release of biologically active materials from polymeric matrices or by cleavage from a polymeric backbone. These systems have been reviewed in some recent books (34,35).

Philosophy of Bioactive Polymers

The elucidation of the biological behavior of compounds can be approached from many points of view. The first of two extremes is to treat the test species with various amounts, etc. of the test material and to observe what occurs either on a gross sense (death) or on a more selective level (cell, organ, tissue location and effect). The other extreme attempts to understand and investigate on a molecular basis using model systems, in situ tests, etc. The former allows a quick assay of general biological activity of the test compound while the latter gives more detailed, but more limited in scope, information about the biological activity of the test compound under more controlled, but typically artificial conditions and extrapolation to real systems is necessary. Both approaches have their advantages and both should be utilized in appropriate situations.

Desirable objectives in the design of chemotherapy drugs include the synthesis of materials which have (a) good specificity, (b) good activity, (c) prolonged activity, and (d) a wide concentration range of biological activity. Thus variability and specificity are key ingredients.

Polymeric drugs allow both variability and specificity to be built into their structure. The placement and chemical environment of the biologically active agent can be varied within the polymer along with introducing segments which can vary the macromolecule's solubility and uptake in select organs and tissues. The molecular shape and size can also be varied to achieve desirable flow and isolation (cell walls acting as barriers, etc.). A similar polymeric design can be envisioned for polymeric herbicides, pesticides, etc.

Complexity of Bioactive Polymers

Natural systems are exceedingly complex and interrelated. As scientists, we become entangled in this web from every side as we attempt to increase our knowledge of thse varied systems which include such diverse areas and entities as heart disease, cancer, nutrition, vitamins, enzymes and nucleic acids woven together into an elaborate network. Slowly we observe that these effects are manifested on a molecular as well as a cellular level, and we observe further that entire systems of organs are frequently directly involved in these interactions. At times it appears nearly impossible to make any simple generalizations in this complex field of biologically active polymers.

Many biological studies are attempted utilizing poorly characterized materials, inadequately understood procedures, etc. In our laboratories we have discovered examples where biological analyses were carried out using supposedly pure small molecules purchased from appropriate chemical and biological supply companies which on further analysis were shown to be mixtures of several compounds. Analyses (such as amino acid analyses) are conducted by the biologist on a "black box" premise and we as chemists often view the world of biological assays also on a "black box" basis. Much of this "black box" treatment is due to an attempt to cope with a complex entity-life in all its varied forms and levels. Even so we must continue the process of lighting candles in these "black boxes" where possible.

This complexity has given birth to and continues to involve new disciplines such as anatomy, biological chemistry, biophysics, bioengineering, microbiology, immunology, virology, physiology, cardiology, pharmacology, toxicology, dermatology, endocrinology, epidemiology, gastroenterology, infectious diseases, neonatology, nephrology, neurology, otolaryngology, pathology, etc. and directly contributes to the interdisciplinary nature of evaluation of the biological activities of compounds.

The use and understanding of macromolecules as biologically active agents compounds this complexity. Polymer purity, molecular weight, molecular weight distribution, configuration, conformation, nature of end groups all may affect the biological activity of the macromolecules. Time takes on a greater meaning for macromolecules: aging, changes in the chemical and physical nature, flow dynamics, solubility, ability to penetrate cellular barriers, etc. all may change with time.

An added problem concerns the languages utilized by the various group disciplines in describing test procedures and reporting biological results. Each of the biologically related disciplines, government, industrial and clinical laboratories have evolved different reporting procedures typically significant for the mode and purpose of the employed tests. To this is often added additional languages according to the particular disease or tested species. While such diversity is natural, and often legitimate to reflect specific test procedures, organisms, etc., it is necessary that such variances do not impede progress, and interaction between different disciplines is essential in order to bridge this gap.

Interdisciplinary Nature

To realize fully the potential of the biological activity of any compound there must be a marriage between chemistry and biology. While the discipline of biochemistry bridges this gap with respect to many topics, it generally does an inadequate job with respect to drugs, and in particular with respect to polymeric biologically active agents. This void is partially being overcome with the emergency of several interdisciplinary doctorate programs. An even more severe problem exists in situations requiring clinical assays where only qualified MD's are permitted to direct (human) clinical programs.

Typically the chemical and biological areas are bridged through combining the efforts of two or more persons working as a team. In all cases it is best for the team members to attempt to learn the language, strategy of one another and for all to contribute to the planning and conducting of the continuing project on an equal basis.

Modes of Bioactivity

A bioactive polymer could interact directly or indirectly with another molecule or an active site to elicit a biological response. (An indirect mode of action would involve a chemical and/or physical interaction with some other chemical species which would then make the polymer biologically active.) The bioactive moiety could be: (a) the entire polymer chain, (b) segments within the polymer chain, (c) oligiomeric degradation products of the polymer chain, or (d) a low molecular weight (monomeric) fragment derived from the polymer. Modes (c) and (d) would be termed as activity via controlled release of the bioactive agent or species while modes (a) and (b) would be a more direct biological activity of the polymer itself. All these modes (a-d) could occur via direct or indirect interactions.

An additional mode utilizes the polymer as a matrix material for the embedding of the biologically active agent. These systems are called embedded or encapsulated systems. In these systems the biologically active agent is embedded within the polymer matrix and the agent is slowly released by a diffusion process or through the decomposition of the polymer within or about the target species. The polymer embedded agent can be either a reservoir system, in which the agent is coated or surrounded by the polymer, or it can be a monolithic system where the drug is distributed fairly uniformly throughout the polymer matrix. Much research has been done on these systems and embedded drugs show great promise in the treatment of numerous medical disorders such as glaucoma and heroin addiction, in birth control, and in the control of bacteria, fungus and insects. The topic of embedded drugs will only be lightly dealt with within the following chapters.

Advantages

Many (potential) advantages exist in the use of biologically active macromolecules compared to the usually low molecular weight agents. These include: (a) sustained activity, (b) greater specificity of action (c) lower toxicity, (d) reduced undesirable side effects, and (e) more direct use of the polymer associated active agent. Part of these advantages stem from the ability to incorporate multiple features into macromolecular chains more readily than into a low molecular weight compound and from the ability to restrict the contact which the polymeric material has towards a test organism more readily than with a low molecular weight compound. Conventional drug delivery involving higher test species, mainly animals, involves ingestion of the drug orally into the digestive tract or injection into the circulatory system. The drug is then transported to the diseased site. Unfortunately, significant portions of the drug are typically excreted prior to this and, in addition, a large amount is transported to other tissues and organs which are not diseased resulting in unwanted side effects. Thus the therapeutic index (ratio of toxic dose to the therapeutic dose) of the drug is typically lowered and the duration of activity low. Polymeric drugs can also be delivered through injection and oral administration and through implantation. Polymeric drugs with limited solubility can be implanted or injected at or near the diseased site and would tend to diffuse only slowly. This reduces the undesired side effects and permits a lengthened duration of activity. Alternatively, a soluble polymeric drug can be taken orally or injected into the circulatory system. Two distinct possibilities exist. First, the polymeric drug could diffuse through the body but due to its size, its passage into certain tissues and organs would be restricted. The second possibility is to tailor the properties of the polymeric drug so it would act as a "guided missile" zeroing in on the desired target of activity (1,11,12,26).

The topic of selective targeting of drugs is currently the weakest link in the overall drug delivery system. Work is involved in the use of coupled drugs (insulin tends to make cell walls more permeable, thus it is being investigated in the delivery of anticancer drugs), embedding in synthetic and natural "homing devices" (lipisome embedded drugs) and in the design of drugs with both the targeting and drug activity built in. This work includes use of polymeric drugs in all three areas. Here we will restrict our comments to the latter approach.

It is reasonable to consider that a polymeric drug can be constructed to contain portions which (a) exhibit the desired pharmacological activity, and (b) direct the polymer to a specific site through solubility, steric, electronic and size tailoring. These special features would enable the polymeric drug to have specific activity, reduced side effects and presumably a higher therapeutic index compared to the analogous low molecular weight drug. Similar features could also be built into other bioactive polymers such as polymeric herbicides, etc.

Finally, the macromolecular nature of the polymeric drug might be an advantage. For instance, one mechanism for the anticancer activity of cis-dichlorodiamminoplatinum II,(C-DDP) is that it binds to the two DNA strands through ligand sites supplied by vacated chloride ions thus rendering the DNA dual strands sterile. For this to occur the "window" is quite small with rigorous steric, reactivity and electronic requirements. For instance, hydrolysis of one of the chloro groups prior to or subsequent to initial complexing of the platinum moiety to one DNA strand would prevent the platinum moiety from fulfilling its task. A polymeric analog would have many more opportunities to render the DNA strand couple inactive.

Conclusions

Bioactive polymers have existed since the creation of life on this earth and exhibit many specific interactions. Synthetic, bioactive polymers are of a much more recent origin but show great potential for highly specific bioactivity in the areas of chemotherapeutic agents, herbicides, pesticides, etc. Future developments in the field of biologically active polymers will probably result in greater control over medication and pest-control agents which will aid greatly in relieving the stress currently being placed on the environment or on the human body by the presently used agents. These advances will result primarily as a consequence of the macromolecular nature ofthese bioactive polymers.

Literature Cited

1. Donaruma, L.G., "Progress in Polymer Science," Vol. 4, 1974, Pergamon Press, New York, p. 1.
2. Gebelein, C.G., Polymer News, 1978, 4, 163.
3. Kuhn, R., Dann, O., Ann. Chem. 1941, 547, 293.
4. Mowry, D.T., Renoll, M., Huber, M.F., J. Am. Chem. Soc., 1946, 68, 1105.
5. Herz, W., Coutney, C.F., J. Am. Chem. Soc., 1954, 76, 576.
6. Puetzer, B., Katz, L., Horwitz, L., J. Am. Chem. Soc., 1952, 74, 4959.
7. Overberger, C.G., Kogon, J. Am. Chem. Soc., 1954, 76, 1879.
8. Overberger, C.G., Michelotti, J. Am. Chem. Soc., 1958, 80, 988.

9. Overberger, C.G., Salamone, J.C., Acc. Chem. Res., 1969, 2, 217.
10. Overberger, C.G., Podsiadly, C.J., Bioorg. Chem., 1974, 3, 35.
11. Shimidzu, T., "Adv. Polymer Sci.", 1977, 23 Springer-Verlag, New York.
12. Pavlisko, J.A., Overberger, C.G. in "Biomedical and Dental Applications of Polymers", Gebelein, C., Koblitz, F.F., Ed., 1981, Plenum, New York, p. 257.
13. Dombroski, J.R., Donaruma, L.G., Razzano, J., J. Medicinal Chem., 1971, 14, 993.
14. Ringsdorf, H., J. Polymer Sci., 1975, Symp. No. 51, 135.
15. Batz, H.-G., Ringsdorf, H., Ritter, H., Makromol. Chem., 1974, 175, 2229.
16. Netter, K.J., Ringsdorf, H., Wilk, H.-C., Makromol. Chem., 1976, 177, 3527.
17. Carraher, C., Dammeier, R., Makromol. Chem., 1970, 135, 107.
18. Millich, F., Carraher, C., U. S. Patent 3,491,061; 1970.
19. Carraher, C., Inorg. Macromol. Revs., 1972, 1, (4), 271, 287.
20. Carraher, C. in "Interfacial Synthesis, Vol. II", Millich, F., Carraher, C., Ed., 1977, Dekker, New York, Chapters 20,21.
21. Subramanian, R.V., Garg, B.K., Corredor, J. in "Organometallic Polymers", Carraher, C., Sheats, J., Pittman, C., Ed., Acadmic, New York, 1978, p. 181.
22. McCormick, C.L., Fooladi, M.M. in "Controlled Release of Bioactive Materials", Baker, R.W., Ed., 1980, Academic, New Yori, p. 317.
23. Takemoto, K., J. Polymer Sci., 1976, Symp. No. 55, 105.
24. Akashi, M., Kita, Y., Inaki, Y., Takemoto, K., Makromol. Chem., 1977, 178, 1211.
25. Pitha, J. in "Biomedical and Dental Applications of Polymers", Gebelein, C.G., Koblitz, F.F., Ed., 1981, Plenum, New York, p. 203.
26. Gebelein, C.G., Morgan, R.M., Glowacky, R., Baig, W. in "Biomedical and dental Applicatoins of Polymers", Gebelein, C.G., Koblitz, F.F., Ed., 1981, Plenum, New York, p. 191.
27. Gebelein, C.G., J. Macromol. Sci.-Chem., 1971, A5, (2), 433.
28. Gebelein, C.G., Baytos, A. in "Reactions on Polymers", Moore, J.A., Ed., 1973, Reidel, Dordrecht, p. 116.
29. Butler, G.B., J. Polymer Sci., 1960, 48, 279.
30. Breslow, D.S., Pure & Appl. Chem., 1976, 46, 103.
31. Schlipkoter, H.W., Brockhaus, A., Klin. Wochenschr., 1961, 39, 1182.
32. Tsuchida, E., Honda, K., Polymer J., 1975, 7, 498.
33. Bayer, E., Holzbach, G., Angew. Chem. Internat. Ed., 1977, 16, 117.
34. Cardarelli, N., "Controlled Release Pesticide Formulation", 1976, CRC Press, Cleveland.
35. Baker, R., "Controlled Release of Bioactive Materials", 1980, Academic, New York.

RECEIVED August 26, 1981.

NONMEDICAL APPLICATIONS

Biological Activities of Metal-Containing Polymers

CHARLES E. CARRAHER, JR., DAVID J. GIRON, DELIE ROSELYN CERUTIS, WAYNE R. BURT, R. S. VENKATACHALAM, TIMOTHY J. GEHRKE, SHUZO TSUJI, and HOWARD S. BLAXALL

Wright State University, Departments of Chemistry and Microbiology and Immunology, Dayton, OH 45435

> Preliminary biological assay results (bacterial and fungal inhibitory test results) for a number of metal and organometallic polymers are reported emphasizing both the variability and constancy of biological responses. For dextran modified with various organostannane halides, results are dependent on the nature of the organostannane but independent of the modification procedure. Also reported are biological assays of uranium containing compounds where the toxicity of the uranium is greatly reduced through complexation of the uranyl ion with a variety of Lewis bases.

The use of metal-containing polymers for applications in medicine has focused on utilization of siloxane polymers as biologically inert building materials. The purposes of this paper are to illustrate a) the variety of potentially available metal-containing condensation polymers which should exhibit activity toward biological organisms and b) the diversity of biological responses already found for selected metal-containing condensation polymers.

Experimental

The metal-containing polymers cited in this paper were synthesized utilizing low temperature condensation processes - namely the interfacial and solution processes. Where possible, syntheses were accomplished using systems that produced polymers within a minute or less reaction time. The combination of mild reaction conditions and short reaction time permits the use of a variety of potentially unstable reactants where thermal or solution induced rearrangements are possible (1).

Three general biological analysis techniques were utilized. These are the slant tube, disk and protein assays. Slant culture assays were conducted using saline, saline-sugar and saline-nutrient broth solutions. The paper disk assays were conducted with the test organism suspended in sterile water. Appropriate dilutions are made in Sabourand's dextrose agar and the solutions poured into Petri plates. The paper disks are soaked in solutions containing the tested material (solutions) or are placed in the Petri plates with specified weights of test material (solid) placed on them. Protein analyses of samples suspended by continuous vibration were carried out as follows. Cold perchloric acid was added to each culture tube after seven days. Pellets, consisting of the test organism, were collected by centrifugation and subsequently suspended in sodium hydroxide. After autoclaving, the solubilized protein was determined using bovine serum albumin, fraction V, at the standard.

Dimethylsulfoxide, DMSO, and DMSO-water are typically employed as the solvent since many of the polymers are soluble in DMSO and DMSO is not appreciably toxic to man or to most of the organisms tested.

Results are averages of at least duplicate examinations.

Results and Discussion

Variety of Available Polymers

The use of metal-containing compounds as medicines is wide spread and ancient including merbromine (mercurochrome;mercury), silver sulfadizine (prophylactic in treatment for severe burns), 4-ureidophenylarsonic acid (therapy of ameblasis, tryparsamide) and antimony dimercaptosuccinate (schistosome). The appearance of metal-containing macromolecules in the human body is extensive including the metals of iron (transferrin and hemoglobin), vanadium (hemovanadin), molybdenum (xanthine oxidase), zinc (carbonic anhydrase) and copper (hepatocuprein).

Objectives in the design of chemotherapy drugs include that the drugs have a) good specificity (differentiation), b) good activity, c) long duration of activity and d) wide concentration range of biological activity. Thus variability and specificity are key objectives. Metal-containing condensation polymers, because of the variety of metal size, chemical environment and electronic structure, offer a wide variability of biological activities and the opportunity for good specificity.

Synthesis of most of the metal-containing condensation polymers can be considered in Lewis acid-base terms as described below.

$$X - A - X \quad + \quad Y - B - Y \quad \rightarrow \quad (A - B)$$

where A has been UO_2, R_3Sb, R_3As, R_3Bi, R_2Pb, R_2Sn, R_2Ge, R_2Si, R_2Ti, R_2Zr, R_2Hf, R_2Mn, $PtCl_2$, MoO_2 and Y-B-Y has been a hydrazine, hydrazide, urea, amine, oxime, amidoxime, diol, dithio, salt of a diacid, sugar, cellulose or if contained on a polymer the Lewis base has contained an alcohol, amine, sulfate, sulfonate, salt of an acid and amidoxime. The Lewis base has also been utilized as a metal carrier with polymers containing the metals Co, Fe, Rh and Ru thus far synthesized. The Lewis base can also contain mixed nucleophiles such as in xanthane dyes, p-aminobenzoic acid (1; Vitamin B_x; antirickettsial), Catalin (2; anticataract), kynurenic acid (3; used in vitamin B deficiency), salicylic acid (4; a nalgesic, antirhenmatic) and Methioprim (5; tumor antagonist).

The polar-condensation linkage is advantageous as a site for breakdown and (potential) controlled release of moieties derived from either or both the comonomers.

Many of the employed metal-containing Lewis acid reactants undergo certain reactions for which analogous products are not formed through attempted condensation with acid chlorides utilizing mild reaction conditions. Thus Cp_2TiCl_2 reacts with salts of dicarborylic acid giving titanium-containing polyesters whereas the reaction with organic diacid chlorides does not yield the analogous condensation product under similar reaction conditions (2).

$$R_2TiCl_2 + {}^-O_2CRCO_2{}^- \rightarrow \{Ti\text{-}O\text{-}C\text{-}R\text{-}C\text{-}O\}$$

$$ClOC\text{-}R'\text{-}COCl + {}^-O_2CRCO_2{}^- \not\rightarrow \{C\text{-}R'\text{-}C\text{-}O\text{-}C\text{-}R\text{-}C\text{-}O\}$$

This is an important consideration if a drug as 1-4, which contains carboxyl groups, is to be delivered utilizing a condensation monomer or polymer.

The chemical environment of the metal and/or any desired drug can be varied as to its intended use. Thus the environment about the polar linkage can be unobstructed and the hydrophobic character minimul (6) or through the use of hydrophobic blocking groups the polar linkage can be sterically hindered and highly hydrophobic (7). Thus both intended preferential biological location of the drug (blood, lipid, etc. compatability) and rate of drug release (from fast, to the polymers itself acting as the delivered agent) can, in theory, be built into a polymer.

The "delivered-agent" can be either the metal or nonmetal moiety or both moieties and the intended purpose beneficial or detrimental or mixed in nature. Thus 8 was synthesized with the purpose of delivering tin, with the diol portion being coincidental. Polymer 9 was synthesized with the purpose of delivering the hormone portion and 10 was synthesized to deliver both the known drug and the essential metal manganese. Regarding the metal, as a general rule Sn, Pb, As, Sb, U can be considered as inhibitory in nature, Mn, Fe as beneficial and Ti, Si etc. as neutral with regard to biological activity.

Thus there exists the potential for "tailormaking" metal-containing biologically active polymers.

Testing Variables

Biological responses are typically dependent on a number of test parameters including state of the tested material (solid or in solution), solvent, organism and test procedure. Since test results are dependent on these variables, it is advantageous to associate testing conditions with an intended application. Much of our research with polymers containing tin, antimony and arsenic is associated with control of mildew and rot for eventual application in topical medications, as thermal insulation and as additives in paints, textiles and paper products.

Figure 1 gives structures of polymers utilized and noted in Table 1. Table 1 contains results as a function of test organism, state and solvent for three tin, two arsenic and four antimony-containing polymers. In the solid state only compounds 1,3 and 8 exhibited good inhibition toward all three organisms. (The organisms are widespread in nature and cause "mildew" and "rot".) Compound 7 inhibited all three test organisms in DMF solution, whereas it was inactive in the solid, whereas compound 3 showed good inhibition in the dry state but only small to no inhibition in HMPA illustrating the variation of biological response with state of the test material. Compound 1 showed good inhibition in DMF but no inhibition when dissolved in HMPA; compound 8 showed good inhibition to the test organisms when dissolved in DMF but no inhibition to the test organisms when dissolved in HMPA illustrating the dependence of biological response on the solvent utilized to deliver the tested compound.

Other instances have been reported relating biological response with state and employed solvent. For instance there exists a direct correlation between the tendency toward inhibition by solid samples of (polyethyleneimine modified through condensation with organostannane halides) and results employing the same solvents dis solved in DMSO (3).

Tin Polydyes

Recently we reported the synthesis of titanium-containing polyetheresters derived from the condensation of Cp_2TiCl_2 with a wide variety of xanthene dyes (4). More recently we effected synthesis of analogous tin containing xanthene dyes through condensation of organostannane dihalides with xanthane salts (5). These products will be called tin polydyes. The xanthene dyes utilized to form the polydyes are typically utilized as cell-coloring agents by

Table 1. Results as a Function of Compound, Test Organisms and Test State

Compound Designation Figure 1.	Amount Tested (mg)	Dry State Organisms				DMF Organisms		
		As. Flavus	Penicillium sp.	As. Fumagatus	As. Flavus	Penicillium sp.	As. Fumagatus	
1	0.4	7	6	N	5	S	4	
2	0.4	N	N	N				
3	1.0	17	25	2				
4	1.0	M	M	M				
5	0.4	S	N	N				
6	0.4	N	S	N				
7	0.4	N	N	N	S	4	S	
8	0.2	6	I	I	5	4	3	
9	0.2	N	N	N				
Control	N	N	S	S	S	S	N	

(continued on next page)

Compound Designation Figure 1.	DMSO Organisms			TEP Organisms			HMPA Organisms		
	As. Flavus	Penicillium sp.	As. Fumagatus	As. Fumagatus	As. Flavus	Penicillium sp.	As. Flavus	Penicillium sp.	As. Fumagatus
1	S	S	S	10			N	N	N
2	N	N	N	I			N	*	N
3							S	N	N
4							S	N	N
5	N	N	N	10			N	*	N
6	N	N	N	10			N	*	N
7	N	N	N				N	N	N
8	N	N	N				N	N	N
9	N	N	N						
10	N	N	I	N			N		

Organisms used — Aspergillus flavus – 240 colony forming units/plate (CFU/plate)
Penicillium species – 500 CFU/plate
Aspergillus fumagatus – 760 CFU/plate

Symbols used — Numbers give the mm of inhibition; I=inhibition, but no clear zone; N=no inhibition; S=slight inhibition, but no clear zone; and * = enhanced growth

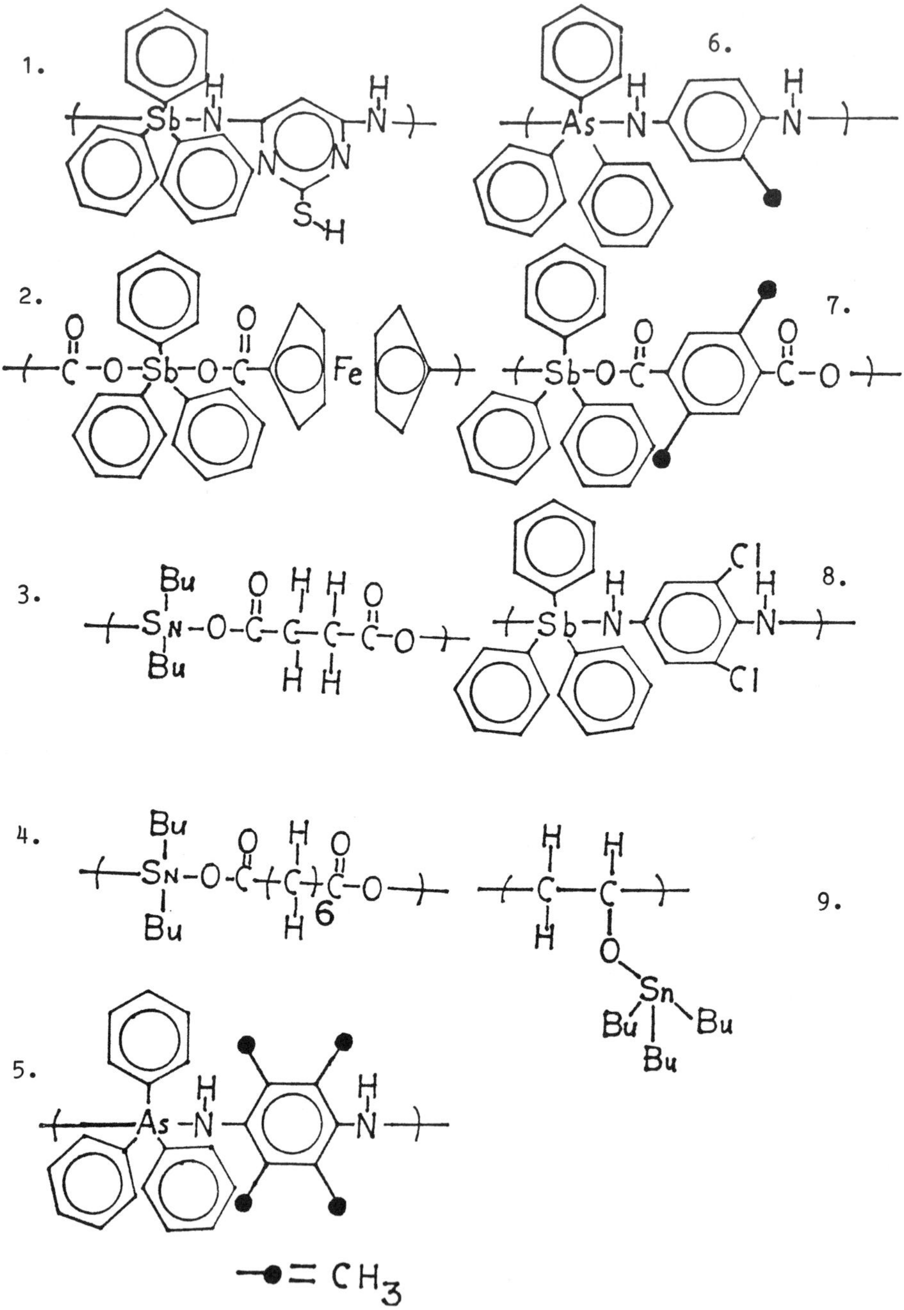

Figure 1. Structures of compounds utilized in biological testing.

microbiologists, biochemists, etc., and as such polymers containing these dyes may be readily accepted by various organisms, effectively delivering the potentially toxic stannane. Further, mercurochrome, a xanthene dye, unlike the other utilized dyes, is toxic to many organisms because of the presence of mercury.

A number of tin polydyes derived from xanthene dyes were tested against the bacterial species Escherichia coli (C600) and Pseudomonas aeruginosa (7430). Figure 2 shows the structures of tested compounds. Results from testing the compounds as solids appear in Table 2. As expected, the greatest toxicity is from the polydye derived from mercurochrome, probably due to at least the toxicity of the dye portion itself. Most of the compounds exhibited some inhibitory nature consistent with the idea that toxic moieties can, in select situations, be "activated" through coupling with moieties already known to be "biologically acceptable."

Table 2. Inhibition as a Function of Organostannane Polyxanthene

Compound Designation From Figure 2	Organism Tested			
	Pseudomonas Aeruginosa		E. Coli	
	Weight Compound (mg)	Inhibition Zone (mm, diameter)	Weight Compound (mg)	Inhibition Zone (mm)
1	10	0	10	1
2	10	0.8	1	0.7
3	4	1.3	10	1.6
4	6	0	6	0
5	10	2	10	2.5

Polysaccharides Containing Tin

Recent efforts have involved condensation of organostannane halides with polysaccharides derived from various sources (6-8). Two polysaccharide sources are cotton and dextran. The inhibition tendencies of the modified polysaccharide are independent of their source but are dependent on the nature of the substituents on the tin such that inhibition generally decreases as the length of the alkyl chain increases, in agreement with the overall general tendencies of organostannanes (Tables 3 and 4; 6-8). The tests were performed using dry samples more closely approximating the conditions under which the polysaccharide compounds may be used.

To assess whether inhibition is due to an inability of the organisms to metabolize the modified polysaccharide or due to the actual toxicity of the products themselves, studies relating the growth in a liquid medium were undertaken. Trichoderma reesei and Chaetominum globosum readily metabolize polysaccharides including cellulose derived from cotton.

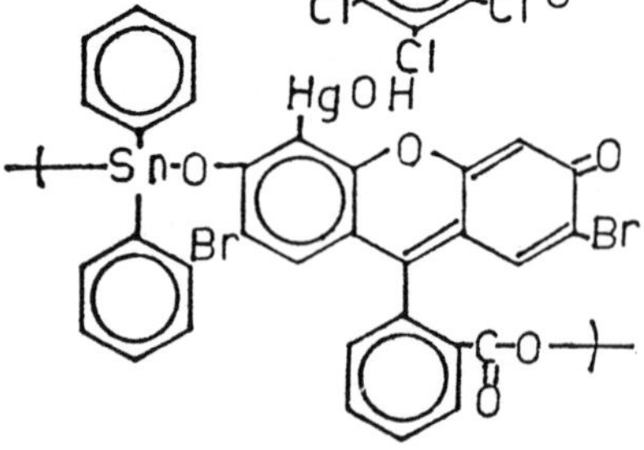

Figure 2. Structures of compounds utilized in biological testing.

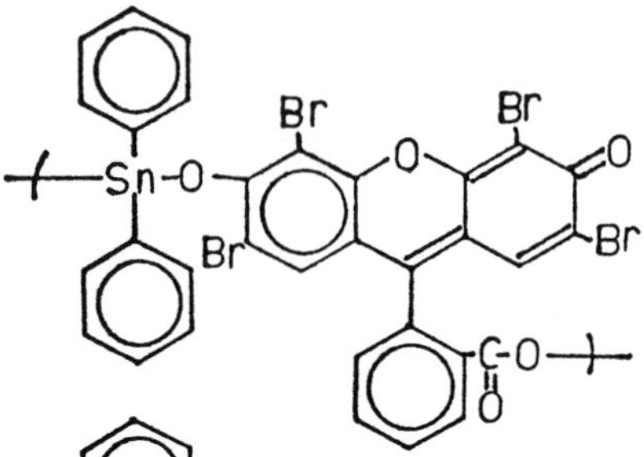

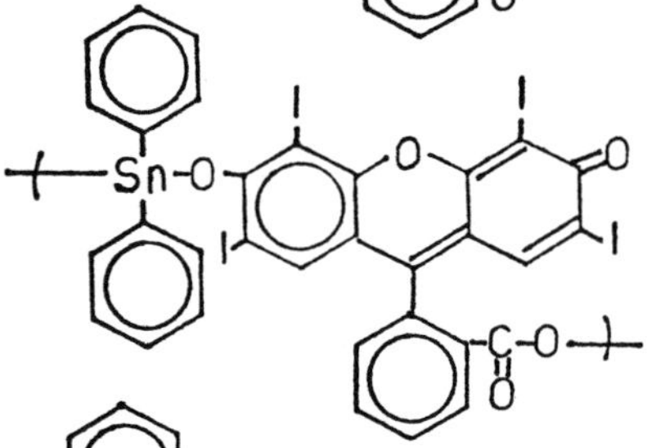

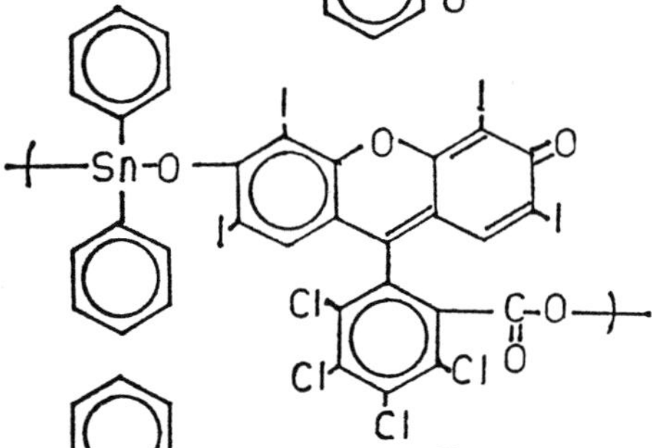

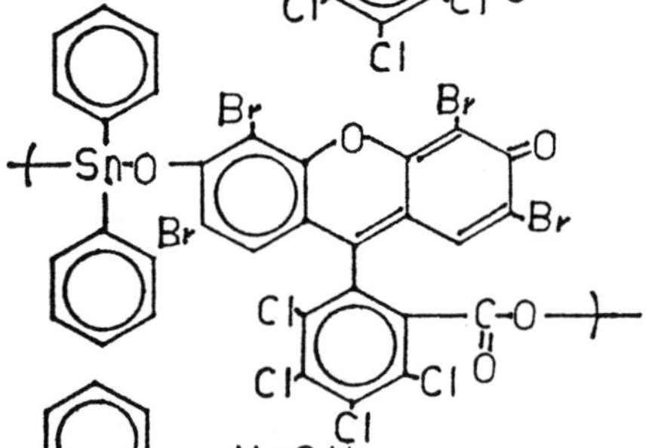

Table 3. Inhibition of Organostannane Modified Polysaccharide Compounds.

Organostannane Dichloride	Tin (%)	Ps. Aeruginosa	E. Coli	A. Flavus	A. Niger	A. Fumagatus
Dimethyltin	8	4	4			
Dipropyltin	21			4	4	3
Dibutyltin	37			4	3	2
Dioctyltin	18	0	0	3	0	0
Dilauryltin	15	0	0			
Dibenzyltin	16	0	4			
Triphenyltin	20	0	0	4	3	3

[a.] 4 = 100%, 3 = 75%, 2 = 50%, 1 = 25%, 0 - 0% inhibition.

Table 4. Growth of Organisms on Organostannane-Containing Cellulose Compounds.

Organostannane Dichloride	Tin (%)	Trichoderma Reesei Growth (ug protein/ml)		Chaetomium Globosum Growth (ug protein/ml)	
		With Dextrose	None	With Dextrose	None
Dipropyltin	21	70	70	20	20
Dibutyltin	40	30	30	40	40
Diactyltin	18	280	40	120	50
Cellulose	-	420	250	300	240

Compared to growth of the fungi in solutions not containing tin modified cellulose, all the compounds inhibit the two fungi (Table 4). Growth inhibition in the dextrose containing media indicates that inhibition is the result of the toxicity of the modified cellulose compounds on the fungi rather than merely an inability of the fungi to degrade the test compounds.

The results are indicative of the applicability of such tin-containing polysaccharides being employed for the retardation of fungi related rot and mildew. Questions such as duration and mechanism of fungal inhibition by these materials have yet to be answered.

Uranyl-Containing Compounds

In select cases the purpose for encapsulating the metal or organometallic in a polymer structure is to diminish biological toxicity. The uranyl ion, in aqueous solution, is reported to be phototoxic to all tested aquatic fungi and bacteria (9,10). The LD_{50} for human beings is probably below 0.1 mg/kg body weight with uranyl ion exhibiting typical heavy metal toxicity (9-11).

Along with developing procedures for the removal, isolation and concentration of the uranyl ion is the concern for developing less toxic forms of the uranyl ion (2,12). Thus preliminary biological examination was undertaken on several representative bacteria. The test samples included uranyl modified PANa, uranyl nitrate hexahydrate, several uranium polyesters and two uranyl polyoximes.

The uranyl nitrate showed a greater zone of inhibition than that of any of the uranyl polyesters and polyoximes and the uranyl modified PANa (Table 5). In fact, all but one of the uranyl polyesters show no inhibition toward the tested species. Thus complexing the uranyl ion with bidentate salts of carboxylic acids decreases the toxicity of the uranyl ion (at least for the tested species). These tests were conducted on solid samples. Further tests are underway.

Table 5. Results of Preliminary Biological Tests as a Function of Uranyl Compound.

		Bacterial Species	
		E.Coli(C600)	Pseudomonas Aeruginosa(7430)
UO_2^{2+} Reacted with salt of	Sample Amount for Test(g)	Diameter of Inhibition(mm)	Diameter of Inhibition(mm)
PAA	.010	1.3	0.6
UNHH (itself)	.010	2.2	2.1
Terephthalic Acid	.010	NI	NI
Succinic Acid	.010	NI	NI
1,1'-Dicarboxylic Ferrocene Dicarboxylic Acid	.010	NI	NI
2,5-Dichlorotetraphthalic Acid	.010	NI	NI
Adipic Acid	.010	NI	NI
Azelaic Acid	.010	1.1	1.0
Azelaic Acid	.0050	1.0	1.7
p-Benzoquinone Dioxime	.010	NI	NI
Terephthalic Dioxime	.010	0.7	0.7
Terphthalic Dioxime	.0050	0.5	0.5

NI = No Inhibition
Diameter of spot maintained to approximately 0.5mm

Literature Cited

1. C. Carraher, Polymer P., 19(2), 100 (1978).
2. C. Carraher, "Interfacial Synthesis, Vol. II. Technology and Applications", (Eds. F. Millich and C. Carraher), Chpt. 21, Dekker, N.Y., 1977.
3. C. Carraher, D. Giron, W. Woelk, J. Schroeder and M. Feddersen, J. Applied Poly. Sci. 23, 1501 (1979).
4. C. Carraher, R. Schwarz, J. Schroeder, M. Schwarz and H.M. Molloy, Org. Coat. Plast. Chem., 43, 798 (1980).
5. C. Carraher, D. Giron, D. Cerutis, T. Gehrke, S. Tsuji, R. Ventatachalam and H. Blaxall, Org. Coat. Plast. Chem., 44, 1 (1981).
6. C. Carraher and T. Gehrke, Org. Coat. Plast. Chem., in press.
7. C. Carraher, J. Schroeder, C. McNeely, D. Giron and J. Workman, Org. Coat. Plast. Chem., 40, 560 (1979).
8. C. Carraher, J. Schroeder, C. McNeely, J. Workman and D. Giron, "Modification of Polymers", (Eds. C. CArraher and M. Tsuda), Chpt. 25, ACS, Washington, D.C., 1980.
9. W. Neely, R. Smith, R. Cody, J. McDuffie and J. Lansden, U.S. Nat. Tech. Inform. Ser., P.B. Rep #220167/1, 1973.
10. "Toxic Substances List - 1974 Eddition", (Edited by H. Christensen and T. Luginbyhl), U.S. Dept. of Health, Education and Welfare, Rockville, Maryland, 1974.
11. "Plutonium and Other Transuranium Elements: Sources, Environmental Distribution and Biomedical Effects", U.S. Atomic Energy Commission, Dec. 1974.
12. C. Carraher and J. Schroeder, Polymer Preprints, 16, 659 (1974) and Polymer Letters, 13, 215 (1975).

RECEIVED July 7, 1981.

Biocidal Activity of Organotin Polymers in Wood

DEBORAH M. ANDERSEN
David W. Taylor Naval Ship R&D Center, Annapolis, MD 20084

J. A. MENDOZA, B. K. GARG, and R. V. SUBRAMANIAN
Washington State University, Department of Materials Science and Engineering, Pullman, WA 99164

The synthesis and preliminary biological assay of organostannane-containing wood is described for the purpose of developing controlled release agents for wood preservation. The organostannane is bonded to the wood using in situ polymerizations. A number of the stannane-containing wood samples show complete resistance to rot and mildew after two years exposure. Further mechanical properties and dimensional stability of the modified wood is also improved.

Trialkyltin compounds have emerged in recent years as broad spectrum toxicants having high toxicity towards marine fouling organisms (1) and wood destroying organisms (2-4) while possessing a tolerable toxicity towards mammals. It has also been shown that trialkyltin compounds pose a minimal hazard to the environment since they eventually degrade to harmless inorganic tin oxides by the action of UV light, microbes, etc. (5,6).

Many controlled release formulations based on tri-n-butyltin groups chemically linked to polymers have been devised in our laboratories and they have performed successfully as antifouling coatings during many months of exposure to marine fouling environment at Miami Beach, Florida (7,8). The effects of chemical structure upon mechanical properties of such polymers (9), as also the copolymerization behavior of organotin vinyl monomers with functional group containing vinyl monomers have been studied (10). Such formulations are effective against Pseudomonas nigrifaciens (marine bacterium), Sarcina lutea (soil bacterium), and Giomeralla cingulata (soil fungus); and the effect of the nature of crosslinks upon leaching rate of the toxin has been determined in our earlier work (11). An investigation of the reactivity of organotin carboxylate polymers has established that the bioactive species released from these polymers in an ocean environment is the trialkyltin chloride (12).

0097-6156/82/0186-0027$5.00/0
© 1982 American Chemical Society

Thus, the aim of our study reported here was to explore ways of adapting the methods of synthesis and application of controlled release organotin antifouling polymers developed in these laboratories to problems of wood preservation (13). The basic approach is to bond the toxic moiety, tributyltin, to a monomer small enough to penetrate wood, and then polymerize this monomer within the wood, i.e., "in situ polymerization" (14). This results in a wood polymer matrix where the accessible void is filled with polymer and, depending upon the monomer, chemically attached to the cellulose structure of wood. Tributyltin compounds, as simple (unbound) additives, are easily leached from wood, and therefore are of limited life expectancy. In the system under study, the treated wood should have a low leach rate because the diffusion of the toxic moiety, bound chemically to the polymer chain, is controlled by the polymer matrix fixed within the wood. This will increase the service life of the wood while ensuring minimal impact on the surrounding environment. Polymer impregnation also has the advantage of increasing the dimensional stability of the wood in water. By decreasing the amount of water which can be absorbed by the wood, the alternating swelling and shrinking of wood in the tidal zone is minimized, and leaching of the preservative is decreased.

Experimental

All treatments to be used in initial field tests were done with standard test coupons provided by the Forest Products Laboratory. These wood specimens are cut from Southern Yellow Pine trees harvested from U.S. Forest Service timber stands and cut into precision 1.5 x 6 x 0.25 inch (3.81 x 15.24 x 0.635 cm) samples with the grain running lengthwise and a ring count of 6 to 10 per inch. By design, each sample therefore has a standard volume of 36.87 cm^3.

All of the samples were treated by one of three different standard procedures for vacuum/pressure impregnation: Full Cell Method, Empty Cell Method (Lowry) or modified Full Cell Method (15). The treatments were applied in a modified pressurized paint tank set up to simulate the procedures used commercially in typical pressure treating cylinders. It was connected to a laboratory vacuum pump in order to pull a vacuum of up to 29 in. Hg (98 kPa), and a compressed air cylinder in order to provide up to 120 psi (827 kPa) air pressure. After treatment, the surface of specimens was wiped free of excess solutions, the samples were air dried overnight, then placed in the circulating air oven and dried to constant weight to remove solvent and moisture before calculating retentions.

The modified Full Cell treatment was employed for impregnating the wood with monomeric materials. This treatment used the treating tank and the vacuum system as a surge tank system. A vacuum-tight glass reaction kettle was used as the treating chamber. It was equipped with a drain in the bottom, which doubled as an inlet for treating solution, and a ground glass jointed addition funnel which has a stopcock connection to the vacuum tubing that was connected to the surge tank. This was used to make sure that the treating solution was drawn up to a level in the addition funnel where there was excess solution in order to keep the wood totally submerged as it soaked up monomer. After the solution was drained, each sample was carefully wiped to remove excess monomer on the surface and cured in an explosion-proof circulating air oven at the desired temperature. An illustrative set of results of in situ polymerization of organotin monomers in Southern Yellow Pine coupons is given in Table 1.

Six test specimens from each treatment were chosen to be placed in the field tests at the Key West Naval Station, Key West, Florida. The test coupons identified with polyethylene tags were bolted to a fiber glass rack. Each rack contained 2 untreated samples which were used as controls monitoring the degree of attack over each 6-month period between inspections. The racks are suspended on nylon line at a depth of ~1 foot (30 cm) below normal low tide. Consequently, the samples are always submerged but are very close to the intertidal zone where Limnoria attack is heaviest, as evidenced by the condition of pilings in that area.

At each semiannual inspection the racks were pulled up, the surface fouling on the coupons was scraped off to expose the surface of the wood specimen and each small panel was inspected carefully and rated by degree of attack on a scale of 0 to 10. A rating of 10 signifies no attack; 9, trace attack; 7, moderate attack; 4, heaving attack; and 0, no structural integrity left, or sample destroyed. Each rack had one control replaced even if neither had a low rating. The inspections take place in June and December, with the heaviest attack taking place between June and December corresponding to the higher water temperatures during this period. This test method corresponds to ASTM D2481, Accelerated Evaluation of Wood Preservatives for Marine Services by Means of Small-Size Specimens.

Results and Discussion

Standard marine creosote, 40% naphthalene creosote, and tributyltin oxide were used to treat test samples by both Full Cell and Empty Cell Methods. These treatments will serve as performance standards in statistical evaluations of the new preservatives.

Two organometallic polymers previously developed, tributyltin methacrylate/methyl methacrylate copolymer (TBTM/MeM) and the tributyltin ester of methyl vinyl ether/maleic anhydride (TBT-MVEt-MAn), were selected for use as wood impregnants. Full Cell and Empty Cell Methods were used to treat wood with TBTM/MeM in "Rule 66" mineral spirits and in combination with P13 creosote. Although the polymers seem to be miscible with the creosote, this combination results in only surface penetration of wood when impregnated. The tributyltin ester of methyl vinyl ether/maleic anhydride in cyclohexanone was used to treat wood samples by both the Full and Empty Cell Methods. Since it was found to be immiscible with creosote, this combination treatment was not used.

TABLE 1. IMPREGNATION OF SOUTHERN YELLOW PINE WOOD WITH TRIBUTYLTIN METHACRYLATE-GLYCIDYL METHACRYLATE MONOMERS AND "IN SITU" COPOLYMERIZATION OF THE SAME

Sample No.	Polymer in Wood, %[a]	Sn in Wood, %[b]
Weight % of Monomer in Reaction Solution = 10%		
7-1a	9.79	1.956
7-1b	10.02	1.997
7-1c	9.95	1.985
7-1d	10.83	2.143
Avg of Samples	10	2.0
Weight % of Monomers in Reaction Solution = 25%		
7-2a	25.64	4.475
7-2b	25.01	4.387
7-2c	26.19	4.551
7-2d	26.22	4.556
Avg of Samples	26	4.5
Weight % of Monomers in Reaction Solution = 40%		
7-3a	37.57	5.989
7-3b	38.79	6.129
7-3c	38.20	6.062
7-3d	37.41	5.970
Avg of Samples	38	6.0

a: $\dfrac{[\text{wt of treated dry wood}] - [\text{wt of untreated dry wood}]}{[\text{wt of untreated dry wood}]} \times 100.$

b: $\dfrac{[\text{wt fraction of Sn in polymer}] [\text{wt of polymer in treated dry wood}] \times 100}{[\text{wt of treated dry wood}]}$

The length of the polymer chain of these two organometallic polymers makes them rather viscous for use as wood impregnants, so it was decided to try impregnating with monomeric materials and polymerizing "in situ". Methyl methacrylate, tributyltin methacrylate, 50/50 weight percent tributyltin methacrylate/ methyl methacrylate with and without a crosslinking agent, and tributyltin methacrylate diluted with mineral spirits were used as impregnating monomers. Benzoyl peroxide was used as the initiator, and all of the monomers were cured in the wood. This method resulted in high retentions of polymer.

The in situ copolymerization of glycidyl methacrylate/ tributyltin methacrylate in Southern Yellow Pine (Table 1) is comparable to that in Grand Fir wood specimens reported earlier (13,14). With Grand Fir, it was found that the ultimate flexural strength, flexural modulus of elasticity, impact strength, and tensile strength of the treated wood sample were all equivalent to or better than untreated samples. Exhaustive solvent extraction indicated a high percentage of the polymer grafted to the wood. The dimensional stability of the wood was improved by treatment, as indicated by the antishrink efficiency after a 48-hour soak in distilled water. Scanning electron microscopy indicated that, in a glycidyl methacrylate treated sample, the polymer's macrodistribution is uniform throughout the sample. Scanning of a tributyltin methacrylate/glycidyl methacrylate sample and a tributyltin methacrylate/maleic anhydride sample showed that in both of these specimens the polymer was uniformly distributed at the surface, but decreased toward the center. This may be a function of the solvent since acetone is a good wood swelling solvent and benzene is not. Electron probe analysis for tin clearly indicated substantial amounts of tin atoms in the cell walls and in the filled cells.

Table 2 presents some exposure data from the Key West Test Site. It is noteworthy that these samples have an unusually large surface-to-volume ratio when compared to standard pilings and therefore represent an extremely severe test condition. This is important in minimizing the length of time a field trial must be monitored to indicate failure or success of a treatment.

After 2 years of exposure of the first ten sets of samples, none of the organotin treatments has shown any attack. Untreated controls have failed completely, demonstrating sample inoculum present for attack. After 18 months of exposure some attack was noted in the marine creosote treated controls. After 24 months, the 23-lb/ft^3 marine creosote had an average rating of 8.3, the 15.85-lb/ft^3 marine creosote had an average rating of 8.4, and the special marine creosote (40% naphthalene) was rated at an average of 9.7. In 36 months, they have been attacked and damaged severely whereas the organotin polymer impregnated systems do not suffer any deterioration.

The results of the field test confirm the evidence obtained in laboratory tests of wood decay (13,14). In these tests it was found that Grand Fir wood specimens treated with organotin polymers were resistant to attack by brown-rot, white-rot and soft-rot fungae, and a marine bacterium (Coniophoro puteana, Polyporous versicolor, Quaetomium globosum and Pseudomonas nigrifaciens respectively).

TABLE 2. FIELD TEST DATA

Sample No.	Treatment	Rating		
		18 months	24 months	36 months
A1	P-13 Marine Grade	10	10	10
A2	Creosote, (23 lb/ft^3)	10	4	–
A3		10	10	4
A4		10	9	7
A5		9	7	–
A6		10	10	4
A13	TBTM/MEM Copolymer,	10	10	10
A14	(13.91 lb/ft^3)	10	10	10
A15		10	10	10
A16		10	10	10
A17		10	10	10
A18		10	10	10
A31	Tributyl Tin Ester of	10	10	10
A32	Methyl Vinyl Ether/	10	10	10
A33	Maleic Anhydride,	10	10	10
A34	(3.28 lb/ft^3)	10	10	10
A35		10	10	10
A36		10	10	10
A37	P-13 Marine Grade	9	7	–
A38	Creosote, (15.85 lb/ft^3)	10	10	–
A39		10	10	10
A40		10	10	10
A41		10	10	–
A42		10	10	0
A43	Special Marine Creosote,	10	9	7
A44	40% Naphthalene,	10	9	4
A45	(27.17 lb/ft^3)	9	10	4
A46		10	10	10
A47		10	10	10
A48		10	10	9
		12 months	18 months	30 months
B44	TBTM/MeM "In situ"	10	10	10
B45	Polymerization	10	10	10
B46	(22.8 lb/ft^3)	10	10	10
B77		10	10	10
B78		10	10	10
B79		10	10	10
B57	TBTM-GMA "In Situ"	10	10	10
B58	Polymerization, 10%	10	10	10
B59	Polymer in Wood	10	10	10
B60		10	10	10

Conclusions

Protection of wood against microbiological decay can be accomplished by impregnation with trailkyltin polymers. Mechanical properties and dimensional stability may also be improved simultaneously.

References

1. A.T. Phillip, Prog. Org. Coat. 2, 159 (1973/74).
2. T. Hof, J. Inst. Wood Sci., 4 (23), 19 (1969).
3. M.P. Levi, J. Inst. Wood Sci., 4 (19), 45 (1969).
4. R. Cockroft, J. Inst. Wood Sci., 6 (6), 2 (1974).
5. A.W. Sheldon, J. Paint Technol. 47 (600), 54 (1975).
6. E.J. Dyckman and J.A. Montemarano, "Antifouling Organometallic Polymers: Environmentally Compatible Materials", NSRDC Report 4136, (Feb. 1974).
7. R.V. Subramanian, B.K. Gard, J. Jakubowski, J. Corredor, J.A. Montemarano, E.C. Fisher, Am. Chem. Soc., Div. Org. Coat. Plast. Chem. Pap., 36 (2), 660 (1976).
8. R.V. Subramanian and B.K. Garg, Proc. Controlled Release Pesticide Symposium, Corvallis, OR, R. Goulding, Ed., (1977), p. 154.
9. R.V. Subramanian and M. Anand, in 'Chemistry and Properties of Crosslinked Polymers', S.S. Labana, Ed., Academic Press, NY (1977), p. 1.
10. R.V. Subramanian, B.K. Garg, and J. Corredor, J. Macromol. Sci. Chem. A11, 1567 (1977).
11. R.V. Subramanian, B.K. Garg and J. Corredor, in Organometallic Polymers, C.E. Carraher, J.E. Sheats and C.U. Pittman, Eds, Academic Press, NY (1978) p. 181.
12. K.N. Somasekharan and R.V. Subramanian, in Modification of Polymers, C.E. Carraher and M. Tsuda, Eds. American Chemical Society, Washington, D.C. (1980) p. 165.
13. Jorge A. Mendoza, "Wood Preservation by in situ polymerization or organotin monomers," Thesis, Department of Materials Science, Washington State University, Pullman, WA (1977).
14. R.V. Subramanian, J.A. Mendoza and B.K. Garg. in Proc. 5th Int. Symp. Controlled Release of Bioactive Materials, Gaithersburg, MD, (1978) p. 6.8.
15. D.M. Andersen, June 1979. Report No DTNSRDC/SME-78/41, David W. Taylor Naval Ship R&D Center, Annapolis, MD.

RECEIVED July 7, 1981.

Polymer-Bound Fungicides for Paints

CHARLES U. PITTMAN, JR., KEVIN R. LAWYER, and K. S. RAMACHANDRAN

University of Alabama, Department of Chemistry, University, AL 35486

In order to prepare coatings which contained chemically bonded fungicides, four vinyl-substituted fungicides were prepared and both copolymerization and terpolymerization studies were carried out with these monomers. The fungicides were pentachlorophenol, 3,4′,5-tribromosalicylanilide, 8-hydroxyquinoline, and 2-benzyl-4-chlorophenol. Three monomers of each were prepared. The acrylates and the chain-extended 2-fungicidalethyl acrylates were prepared of each fungicide in addition to the vinyl ether of each fungicide. The acrylates and chain-extended (2-fungicidalethyl) acrylates were copolymerized with acrylic monomers and stable terpolymer latexes were prepared from (1) pentachlorophenyl acrylate (2) 3,4′,5-tribromosalicylanilide acrylate and (3) 2-pentachlorophenylethyl acrylate with the comonomers methyl methacrylate and n-butyl acrylate. Terpolymer latices of vinyl acetate, 2-ethylhexyl acrylate, and 3,4′,5-tribromosalicylanilide acrylate were made. Copolymers and latexes containing pentachlorophenyl and 3,4′,5-tribromosalicylanilide acrylate were tested against <u>Auerobasidium pullulans</u> in agar dish accelerated growth studies and were found to prevent growth. Several polymers did not exhibit a zone of inhibition around the treated sample as did the free fungicides which could migrate into the agar medium. However, when the polymers were first irradiated by UV light, a zone of inhibition was found indicating photohydrolysis occurred to free the fungicide. This represents a novel mechanism for the controlled release of fungicide.

The long term protection of paint films against mildew defacement is, as yet, an unachieved goal. The problem has received increased attention from the Paint Research Institute since 1971.[1] A variety of fungicides

0097-6156/82/0186-0035$5.00/0
© 1982 American Chemical Society

have been used as additives to paints, but several problems are encountered when fungicides are simply dispersed into the paint. Since films are thin and have high surface areas, water leaching, evaporation, or migration of the fungicide occur. When the fungicide concentration drops below a critical level, mildew may begin to grow.[2] If a fungicide were chemically bound to the polymer backbone of the film, a number of advantages might be realized.[3] Leaching of the fungicide should be retarded. Vaporization might be eliminated. These differences could lead to longer mildew resistant lifetimes of the film. Solubility problems would be eliminated. Biocides which have high water solubilities or high vapor pressures now could be candidates for use in paints whereas they could not have been considered previously. Finally, the toxicity of the paint (to humans) might be reduced. However, polymer-bound biocides may have reduced effectiveness and this point needs to be experimentally tested.

In designing mildewcidal coating polymers, the fungicides can be bound to the polymer in a permanent fashion or in a way that the fungicide may be slowly released. Chemical anchoring of biocides in paints was the subject of a detailed PRI-supported literature search,[4] but little research has appeared on this topic outside of the area of undersea marine coatings where considerable interest existed.[5] Pittman et.al. first chose to prepare polymers where the biocidal functions were attached by ester and amide groups, because it was envisioned that these could be cleaved in the environment created by developing colonies of Aureobasidium pullulans, the major paint defacing organism,[6] and both preinvading and coinvading microorganisms which accompany these colonies. The broad spectrum biocide, pentachlorophenol, was converted to its acrylate and copolymerized with vinyl acetate or ethyl acrylate to provide polymers where ester group cleavage would release pentachlorophenol.[7] These polymers proved to resist the growth of Pseudomonas sp., Aspergillus sp., Alternaria sp. and A. pullulans in petri dish tests.[7] Unlike polymers to which pentachlorophenol was simply added, films of the biocidal copolymers did not exhibit zones of growth inhibition adjacent to the films.[3,7]

To further test the merit of the polymer-anchoring concept, we have synthesized a series of fungicidal monomers where the active agent is attached by both hydrolyzable (ester) groups and nonhydrolyzable (ether) groups. Co- and terpolymerization of these monomers into model paint binderlike polymers was completed. In addition, some stable paint-like latices were prepared. This synthetic program was undertaken to provide reasonably "paint-like" biocidal polymers for both laboratory and outdoor fence testing of the mildew resistance of coating films. Herein, we describe a portion of the synthetic work and some test results.

Fungicidal Monomer Synthesis

Acrylate derivatives were prepared of the known fungicides(1) pentachlorophenol (3) 3,4,5-tribromosalicylanilide (5) 8-hydroxyquinoline and (7) 2-benzyl-4-chlorophenol using acryloyl chloride and amine bases. These are outlined in Scheme I. These acrylates, (1), (3), (5) and (7) were then used to prepare acrylic paint binder polymers. Since each of these fungicidal acrylates are phenolic esters, they are readily hydrolyzed. The various phenoxides function as good leaving groups. Resonance stabilization of charge into the phenyl ring stabilizes the phenoxy anion making it a better leaving group in ester hydrolysis. Another group of chain-extended acrylates was prepared by introducing an ethyl group between the phenol and ester group. The resulting chain-extended acrylates will not hydrolyze as rapidly because they are alkyl esters. The alkoxide anion leaving group is less stable than a phenoxide and therefore is a poorer leaving group. Thus, in addition to the acrylates, the chain-extended acrylates, (2), (4), (6) and (8), of each of the four fungicides was prepared. The acrylate and chain-extended acrylate syntheses are outlined in Scheme I.

The acrylates (1), (3), (5), and (7) were each prepared using acryloyl chloride under normal Shotten-Baumann conditions. The syntheses of chain-extended acrylates (2), (6), and (8) were accomplished by reacting the respective fungicidal phenoxide anion with 2-chloroethanol to generate the corresponding 2-phenoxyethanol. Treatment of these 2-substituted ethanol derivatives with acryloyl chloride and triethyl amine gave the chain-extended acrylates. This route was unsuccessful, however, starting with 3,4,5-tribromosalicylanilide because the displacment of chloride from 2-chloroethanol failed. The oxygen of 3,4,5-tribromosalicylanilide is severly hindered by the ortho bromo and amide functions which sterically retard the S_N-2 displacement. The same reaction on 2-iodoethanol proceeded in poor yield (25%) giving the 2-substituted ethanol which was then esterified with acryloyl chloride to give (4). The direct displacement of tosylate from 2-tosylethyl acrylate resulted in only an 8% yield of chain-extended acrylate, (4). The structure of monomers (1)-(8) were confirmed by infrared and nuclear magnetic resonance spectroscopy.

Vinyl ethers of the fungicidal monomers were also prepared since an ether linkage connecting a fungicide to a polymer would be highly resistant to hydrolysis. Thus, vinyl ether polymers were a synthetic goal in order to determine if the polymers, themselves, could be biocidal since release of the biocide could not occur. Three general synthetic routes were examined. First, the 2-bromoethyl ethers were made of the fungicidal phenols from 1,2-dibromoethane followed by base-catalyzed dehydrohalogenation. Secondly, based on literature precedent,[8] the phenol was treated

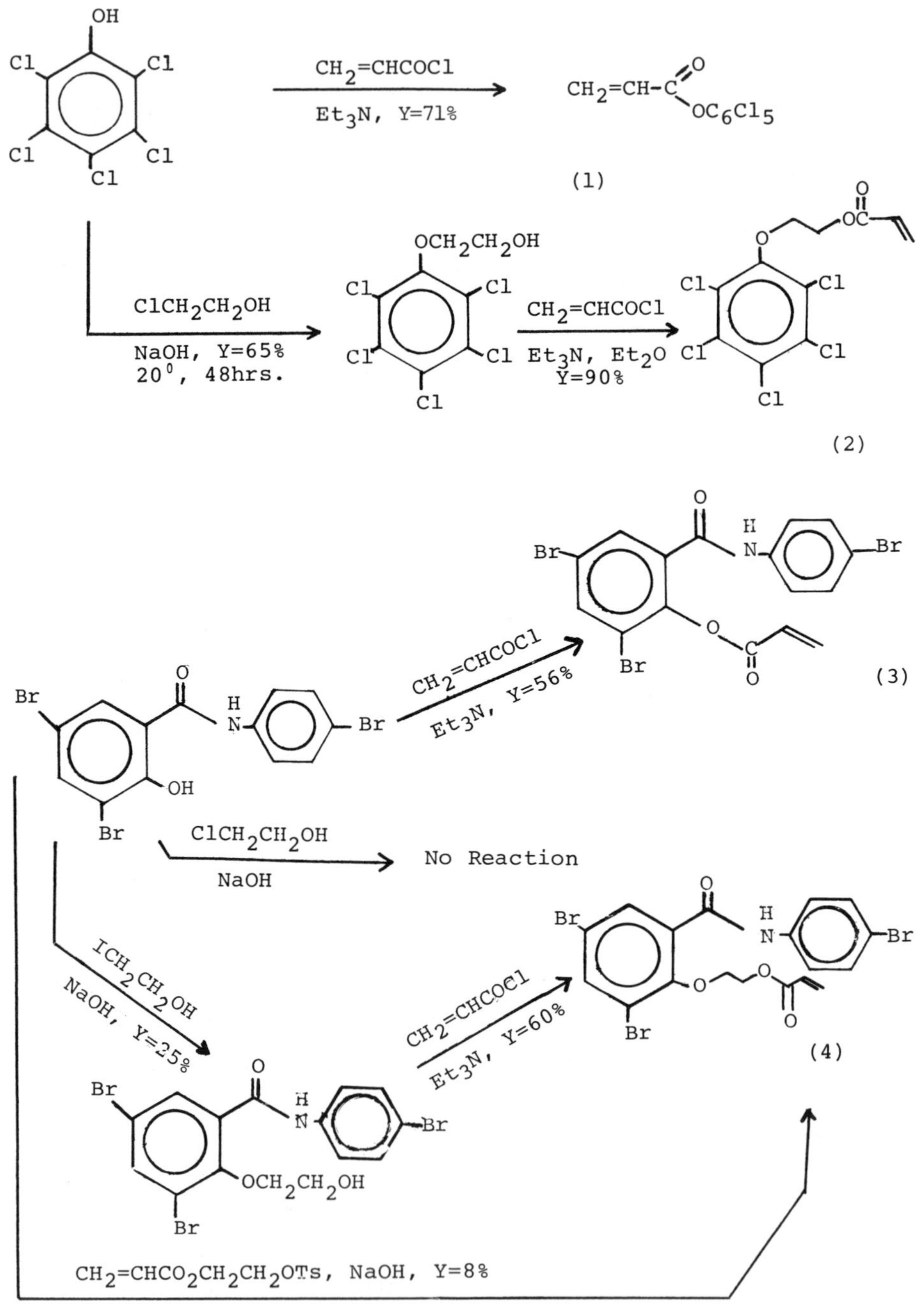

Scheme I. (Part 1)

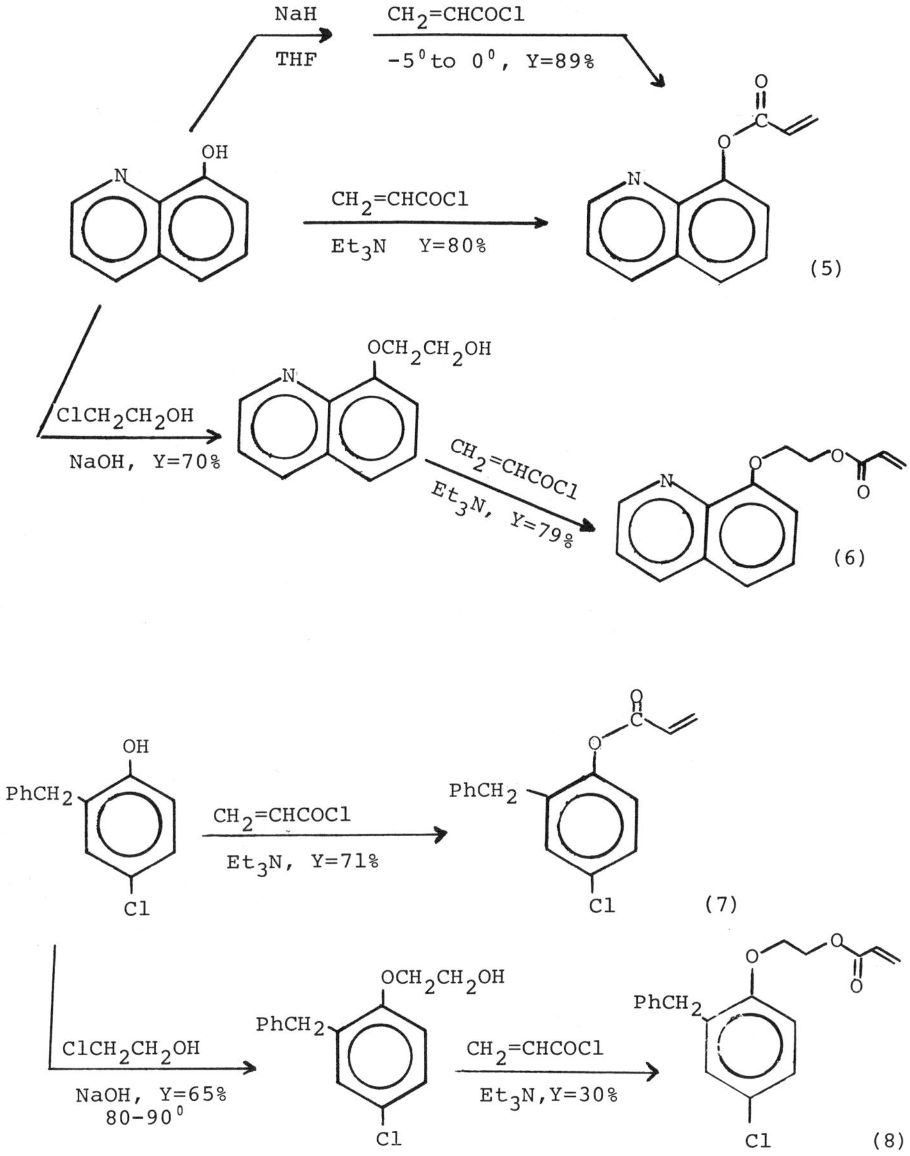

Scheme I. (Part 2)

with vinyl acetate in the presence of mercuric acetate to give the vinyl ether in an addition-elimination sequence. A third route was based on 1,2-vinylbis(triphenylphosphonium)dibromide. Reaction of the phenol with this salt in chloroform and triethylamine, followed by treatment with aqueous NaOH, gave the vinyl ethers of pentachlorophenol and 2-benzyl-4-chlorophenol. This method and the vinyl acetate route were not successful in giving the vinyl ether of the highly hindered phenol, tribromosalicylanilide. 8-hydroxyquinoline was converted to its vinyl ether by treatment with acetylene in base at elevated temperatures and pressures. The routes employed are summarized in Scheme II.

Polymerizations

The fungicidal acrylates, (1), (3), (5), and (7), were solution homopolymerized using radical initiation (AIBN) and copolymerized with hexyl methacrylate in emulsion reactions (sodium lauryl sulfate, $K_2S_2O_8$, $60°$). Similar studies were performed on the chain-extended acrylates, (2), (4), (6), and (8). Emulsion copolymers of (7) and (8) and homopolymers of (7) were of low molecular weight, presumably due to chain-transfer caused by the benzylic hydrogens. The fungicidal monomers give homopolymers with high T_g values. Coatings with fungicidal monomers therefore must have one comonomer which would lower T_g. Thus, n-butyl acrylate, n-hexyl acrylate, or 2-ethylhexyl acrylate were employed in copolymers and terpolymers giving latices which serve as binders in test coating systems.

Pentachlorophenyl acrylate, (1), was terpolymerized with methyl methacrylate (MMA) and n-butyl acrylate (nBA) (Scheme III) to give a latex containing 53% solids and a pH of 4.7 which was adjusted to 6.8 by adding aqueous NaOH. The latex was stable up to pH = 10. A small aliquot was coagulated and the resulting polymer purified. Its intrinsic viscosity was $3.1^{d\ell}/g$ and analysis indicated 2 mole percent (1), 58% MMA and 40% nBA. Similar terpolymer latices were prepared from acrylates (2) and (3) (Scheme III). Another terpolymer latex made from (3), vinyl acetate, and 2-ethylhexyl acrylate contained 54% solids. These latices and their compositions are summarized in the Table 1 and a sample experimental procedure is given in the experimental section.

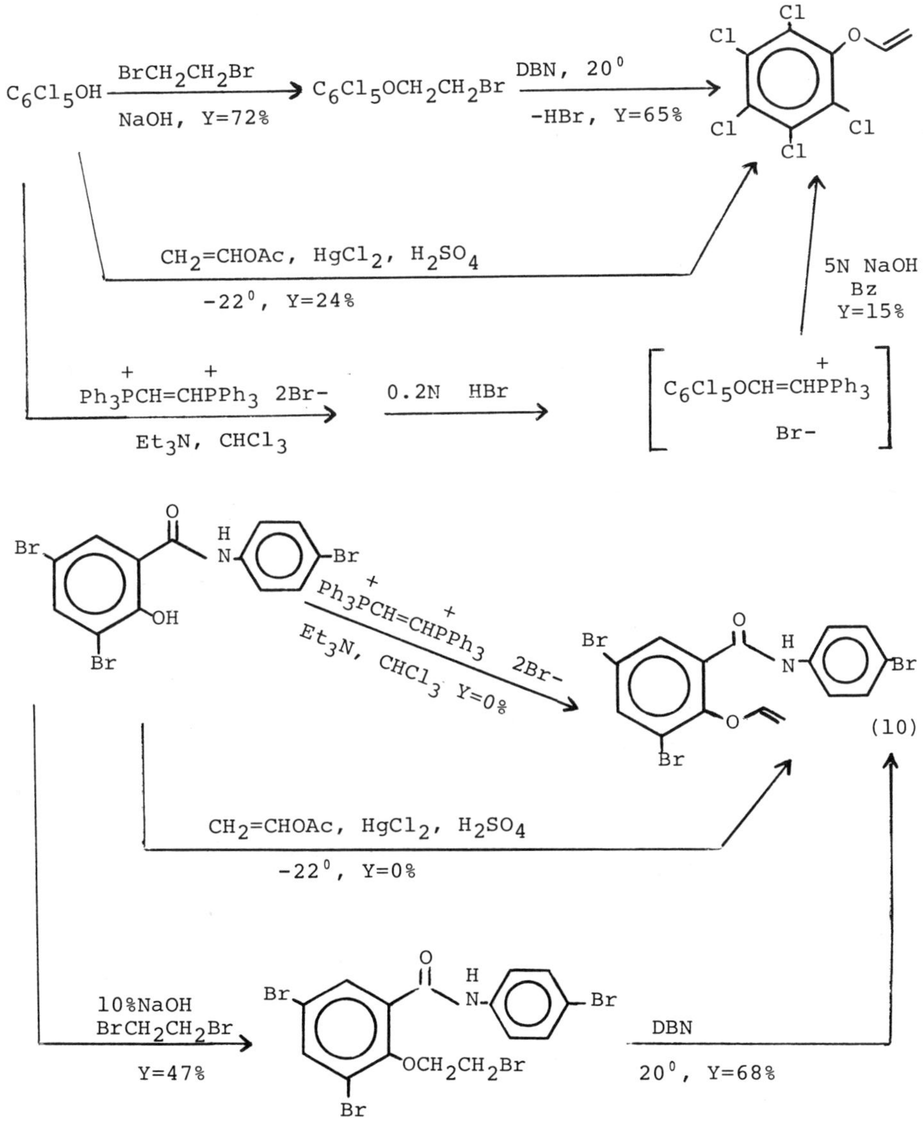

Scheme II. (Part 1)

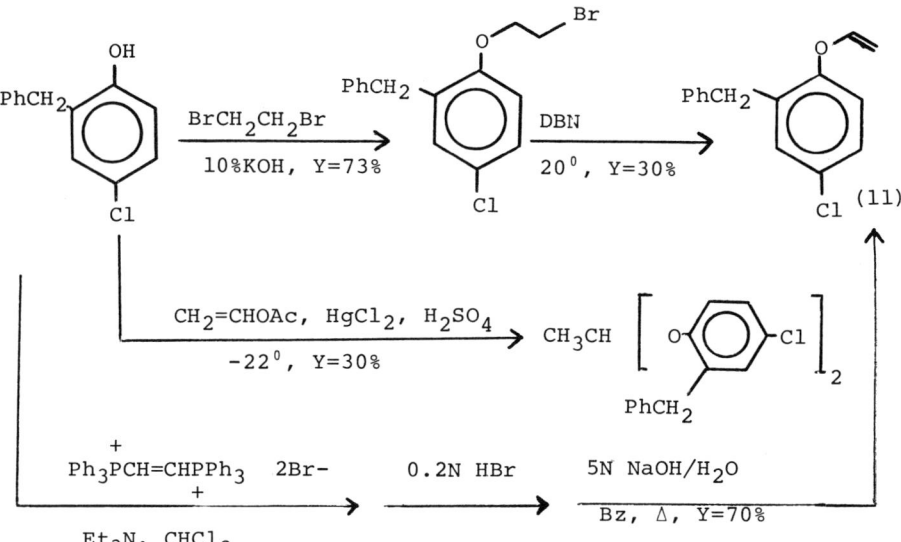

Scheme II. (Part 2)

Table 1. Composition of Stable Latices of Model Paint Binders Containing Chemically Bound Fungicides.[a]

Number	Monomers (mole % in Polymer)[b]	Solids (% wt. of latex)	$\|\eta\|$ [c] dl/g
1	1(2), MMA(58), nBA(40)	49.7	3.1
2	1(5), MMA(64), nBA(32)	40.0	2.3
3	2(11.4), MMA(6), nBA(82.6)	54.2	3.5
4	3(2), MMA(47), nBA(51)	48.3	3.1
5	3(5.8), MMA(43), nBA(51.2)	40.0	2.8
6	3(1.4), VA(87.4), EHA(11.2)	54.0	1.3

[a] These latices were prepared as described in the text. MMA = methyl methacrylate, nBA = n-butyl acrylate, VA = vinyl acetate, EHA = 2-ethylhexyl acrylate.
[b] The numbers in brackets refer to the mole % of the monomers in the binder polymer isolated from the lattices and purified. These were determined by elemental analysis.
[c] The intrinsic viscosities are those of the binder polymer after coagulation, isolation, and purification by two reprecipitation steps.

A variety of petri dish and coated wood accelerated growth tests have been performed with both copolymers and latexes. The results may be generalized as follows. Copolymers of the fungicidal monomers (1) and (3) prevent the growth of mildew without producing zones of inhibition which reach out into the nutrient medium. This was also true of coatings prepared from latex samples 1, 4, and 5 (Table I). This may be contrasted to the free fungicides, pentachlorophenol and 3,4,5-tribromosalicylanilide, which do give zones of inhibition when simply blended (or dissolved) into the coating polymer.

The concentration of fungicide in the coating polymer often must be higher when the fungicide is polymer-bound to prevent mildew growth. This is illustrated for pentachlorophenyl acrylate, (1), and its homopolymer ($\bar{M}n$ = 10,000) in Table II. When tested against Escherchia Coli, Staphylococcus Aspergillus niger, Penicillium citrinum, and Saccharomyces cerevisiae in agar dish tests, to determine the minimum inhibitory concentration (MIC), the homopolymer was less active than monomer (1) when the organisms were E. coli and Saccharomyces cerevisiae. The polymer was of comparable activity for the other organisms in the test. When tested after UV irradiation, coatings prepared from latex numbers 1, and 4 did exhibit zones of inhibition in accelearated growth tests against Auerobasidium pullulans. This illustrated that photohydrolysis was occurring which caused saponification of the ester group, thereby freeing the fungicidal phenol.

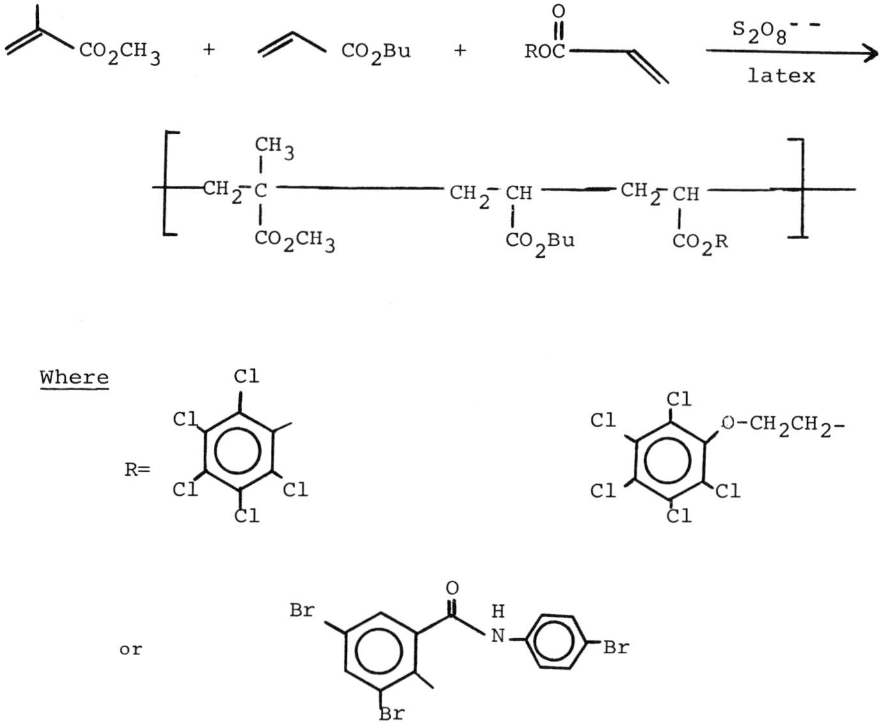
CO2CH3
CO2Bu
ROC
O
S2O8- -
latex
CH3
CH2-C
CO2CH3
CH2-CH
CO2Bu
CH2-CH
CO2R
Where
R=
Cl
Cl
Cl
Cl
Cl
Cl
Cl
Cl
Cl
Cl
O-CH2CH2-
or
Br
O
H
N
Br
Br

Scheme III.

Several copolymers (1)/HMA, (3)/HMA for example) were formulated into test alkyl paint formulations which were painted onto cedar panels for outside evaluation. Latices and copolymers are currently undergoing accelerated growth testing (Zabel, Syracuse University) and they are being formulated into a standard paint (Singer, Troy Chemical Co.) for testing on outdoor fences (Matthews, duPont). Attempts are underway to incorporate vinyl ether into suitable binder polymers. Results of the biological testing will be published later.

Homopolymers and both ethylacrylate and vinylacetate copolymers of 8-hydroxyquinolyl acrylate, (5), gave large zones of inhibition in agar dish tests due to rapid self-hydrolysis from the polymers. Thus, the polymer systems are not suitable for slow release. Tests are underway on polymers of the extended chain analog, (6).

Table II. Minimum Inhibitory Concentrations of Pentachlorophenyl Acrylate, (1), and its Homopolymer Toward Microogranisms Using an Agar Dilution Method. (- means no growth; + means growth).

Compound		MIC					
		1000	500	250	100	50	10
(1)	E. Coli	-	-	-	-	+	+
	Staphylococcus aureus	-	-	-	-	+	+
	Asperigillus niger	-	-	-	+	+	+
	Pencilillium citrinum	-	-	-	+	+	+
	Saccharomyces cerevisiae	-	-	-	-	+	+
homo polymer of (1) M_n=10,000	E. Coli	+	+	+	+	+	+
	Staphylococcus aureus	-	-	-	-	+	+
	aspergillus niger	-	-	-	+	+	+
	Penicillium citrinum	-	-	-	+	+	+
	Saccharomyces cerevisiae	-	-	-	+	+	+

Experimental Section

The synthesis of monomers (1) and (5) have previously appeared. Thus, they will not be further described.

2-Pentachlorophenoxyethanol. 26.6 g (0.1 mole) of sublimed pentachlorophenol was dissolved in water (150 ml) containing NaOH (6 g, 0.15 mole) by warming. After cooling this clear solution to room temp., 2-chloroethanol (12 g, 0.15 mole) 2-chloroethanol in 25 mls of water was added slowly (over 3 hrs) with mechanical stirring. After 10 hrs of stirring, the solution was heated to 80°C for 2 hrs. A copious white precipitate appeared which after cooling was filtered, dissolved in methylene chloride, and washed three times with NaOH (0.5 N) followed by water. The solvent was removed under reduced pressure giving a

white crystalline product which was crystallized from methanol/
water or diethyl ether/pet. ether 20/80 (60%) yield 18 g mp
91°-92° C, = 30.95, H = 1.61 found: 31.03, H = 1.63. IR and nmr
spectra agreed with the structure.

2-Pentachlorophenoxethyl Acrylate, (2). 2-Pentachloro-
phenoxyethanol (16.5 g   0.0532 mole) was dissolved in dry methylene chloride (100 mls) and triethyl amine (6.5 g) was added.
The solution was cooled in ice bath to (0-5°) to which freshly
distilled acrylolyl chloride (5.6 g 0.061 mole) in 20 ml dry
methylene chloride was added under dry conditions. The solution
was allowed to thaw overnight and worked up by addition of saturated solution of sodium carbonate and then by extraction in
methylene chloride. On removal of solvent an oily residue
remained. This was crystallized from methanol/diethylether
(90/10) as light yellow fluffy crystals. M.P. 69-70°C yield 17
gms (90%).

Acryloyloxy-3-4',5-Tribromosalicylanilide, (3). 3,4',5-
Tribromosalicyclicylanilide (30.0 g, 66.7 mmol) and triethylamine
(7.1 g, 70.3 mmol) were dissolved in benzene (100 ml) and cooled
to 5°C. Acryloyl chloride (6.3 g, 69.6 mmol) was added dropwise
with stirring and triethylamine hydrochloride precipitated. The
mixture was warmed to 20°C for 3 hrs., filtered from the triethylamine hydrochloride, washed twice with water, and dried over
anhydrous $MgSO_4$. Removal of the benzene in vacuo gave a white
solid which recrystallized from a 4:1 dioxane/water mixture to
give 18.9 g (53.4%) of 3 mp 149.5 - 151°C. Analysis: Calcd.;
C 38.10; H 1.98; N 2.78; Found: C 38.21; H 1.98; N 2.78; IR
(KBr; cm^{-1}); 3300 (NH), 1740 (CO, ester), 1660, 1630 (CO, amide),
1530, 1390, 860. nmr ($CDCl_3$; ppm): 7.4 - 8.2 (m, aryl and N$\underline{H}$, 7
H), 6.1 (m, 3H, vinyl). The multiplet at 6.1 - 6.7 exhibits an
ABC pattern.

2-(2-Hydroxyethoxy)-3,4',5-Tribromosalicylanilide. Method A:
A solution of 3,4',5-tribromosalicylanilide (4.0 g, 8.8 mmol) in
10% aqueous NaOH 96 ml) was heated to refluxing and 2-chloro-
ethanol (12. g, 15.0 mmol) was added dropwise. After 11 hrs.
the mixture was cooled and extracted with ether and the solution
dried with anhydrous $MgSO_4$. Removal of the ether in vacuo gave
0.9 g of a brown solid which proved to be 3,4',5-tribromosalicyl-
anilide. The aqueous layer was acidified and extracted with
ether. Removal of the ether in vacuo gave back unreacted
3,4',5-tribromosalicylanilide quantitatively.
    Method B: 3,4',5-Tribromosalicylanilide (36.7 g, 81.6 mmol)
was dissolved in a solution of ethanol (50 ml), dimethylformamide (200 ml) and water (50 ml) containing KOH (5.9 g, 105.2
mmol). The solution was added to 2-iodethanol (15.1 g, 87.8
mmol) and the reaction was refluxed for 3 days, filtered and the
filtrate and precipitate were extracted with ether. The combined

extracts were dried with anhydrous MgSO$_4$ and cooled to obtain white crystals of the title compound. Yield: 8.8 g (21.5%). mp 161.8 - 163.0°C. Analysis: Calcd.: C 36.5; H 2.40; N 2.80; Found: C 36.52; H 2.27; N 2.86. IR (KBr; cm^{-1}); 3500 (OH), 3300 (NH), 2900 (aliph. CH), 1650 and 1580 (CO), 1530, 1240 (COC), 1080 (COC). nmr (CDCl$_3$-DMSO-d$_6$ ppm): 7.8 (4H, amide ring), 7.5 (q, 2H, acid ring), 10.3 (broad S, 1H, N$\underline{H}$), 4.1 (m, 2H, OC$\underline{H}_2$), 3.8 (m, 2H, BrC$\underline{H}_2$), 4.8 broad S, O$\underline{H}$).

2-(2-Acryloyloxethoxy)-3,4;5-Tribromosalicylanilide, (4).
Method A: The triethylamine salt of 3,4;5-tribromosalicylanilide (7.3 g, 13.2 mmol) was dissolved in dimethylformamide (10 ml) and added to a solution of 2-(p-toluenesulfonyl)ethyl acrylate (5.5 g, 20.4 mmol) in benzene (20 ml). The mixture was heated to 93°C for 1 hr. and then cooled to 20°C for 16 hrs. The reaction was filtered and water (30 ml) added to the filtrate. The filtrate was extracted with ether and the extract dried with anhydrous MgSO$_4$. Removal of the ether in vacuo gave a brown solid and a brown liquid (unreacted 2-(p-toluenesulfonyl)ethyl acrylate). The solid was recrystallized from chloroform to obtain 0.4 g (5.6%) of (4). mp 153-154°C. Analysis: Calcd.: C 39.42; H 2.55; N 2.55; Found: C 39.53; H 2.61; N 2.52. IR (KBr; cm^{-1}): 3500 (NH), 1740 (CO ester), 1650 and 1580 and (CO amide), 1250, 1120, 1070. nmr (DMSO-d$_6$; ppm): 8.0 (d, 1H, acid ring), 7.7 (d, 1H, acid ring), 7.5 (d, 4H, anilide ring), 7.4 (s, N$\underline{H}$), 5.8 (m, 3H, vinylic, ABC pattern), 4.3 (m, 4H, the ethylenic hydrogens show an A$_2$B$_2$ pattern).
Method B: 3,4;5-Tribromosalicylanilide (13.4 g, 29.7 mmol) was dissolved in a mixture of dimethylformamide (20 ml) and ethanol (50 ml) and the solution was added to 9.9% aqueous NaOH (11.9 g, 29.7 mmol of NaOH). The resulting solution was heated to refluxing and 2-bromoethyl acrylate (5.3 g, 29.7 mmol) was added dropwise. After 8 hrs. the reaction was cooled and (4) crystallized out of the solution. The (4) was recrystallized from chloroform to give 1.2 g (8.1%).

2-(8-Quinolinyloxy)ethanol. 8-Hydroxyquinoline (14.5 g, 100 mmol) was dissolved by warming in 200 ml of water containing NaOH (4 gm). After cooling to room temperature, 2-chloroethanol (12 g, 150 mmol) was added to the solution over 5 hrs. The mixture was stirred for 49 hrs. at 20°C and a precipitate was filtered. The precipitate was water washed, dried in vacuo and recrystallized from benzene/ether/pet.ether 50/20/30. Yield 14.5 g (70%). mp 84-85°C (lit. 83-84°). A pure sample was prepared by chromatography on an alumina column using chloroform as the eluent. Analysis (calculated for C$_{11}$H$_{11}$NO$_2$·H$_2$O): Calcd.: C 63.77; H 6.29; N 6.76; Found: C 63.49; H 6.26; N 6.66. IR (KBr; cm^{-1}): 3400 ()H), 1310 (CO alcohol), 1260 (COC), 1110 (CO alcohol), 1070 (COC). nmr (CDCl$_3$; ppm): 8.8 (m, 1H, aryl), 7.9 (m, 1H, aryl), 7.2 (m, 3H, aryl), 6.9 (q, 1H, aryl),

5.1 (broad S, O$\underline{\text{H}}$), 4.1 (m, 4H, ethylenic showing A_2B_2 pattern). The hydrogens of the nitrogen ring system show an AMX pattern where H^1 is a doublet of doublets at 8.8, H^3 is a quartet at 6.9, and H^2 is a doublet of doublets at 7.9.

2.17 **2-(8-Quinolinyloxy)ethyl Acrylate, (6).** 2-(quinolinyloxy)ethanol (9.7 g, 51.3 mmol) and triethylamine (5.2 g, 51.3 mmol) were dissolved in benzene (117 ml) and cooled to 5°C. Acrylolyl chloride (7 g, 63.0 mmol) was added dropwise and, after the addition was complete, the mixture was warmed to 20°C for 12 hrs. The benzene solution was washed twice with 5% aqueous $NaHCO_3$ (100 ml) and three times with water (100 ml). The benzene solution was dried with anhydrous $MgSO_4$ and the benzene removed in vacuo to give 9.9 g of a brown oil which recrystallized from carbon tetrachloride to obtain 9.2 g (73.8%) of mp 96.5-97.0°C. Analysis: Calcd.: C 69.14; H 5.35; N 5.76; Found: C 68.95; H 5.44; N 5.71. IR (KBr; cm^{-1}): 3050 (aryl CH), 2850 (aliph. CH), 1750 (C=O), 1630 (vinyl C=C), 1600 (aryl C=C), 1200 (ArOC), 1100 (COC), 1040 (ArOC), 980, 930. nmr ($CDCl_3$, ppm): 8.8 (m, 1H, aryl), 8.0 (m, 1H, aryl), 7.2 (m, 4H, aryl), 6.0 (m, 3H, vinyl: ABC pattern), 4.5 (m, 4H, ethylenic, A_2B_2 pattern).

o-Benzyl-p-chlorophenyl Acrylate, (7). o-Benzyl-p-chlorophenol(29.7 g, 126 mmol) and triethylamine (12.7 g, 126 mmol) were dissolved in benzene (100 ml) and cooled to 5°C. Acryloyl chloride (11.4 g, 126 mmol) was added dropwise with vigorous stirring over a period of 45 min. and then the reaction was warmed to 20°C for 3 hrs. The reaction mixture was washed once with 10% aqueous $NaHCO_3$ and twice with water, dried with anhydrous $MgSO_4$. Removal of the benzene in vacuo gave a brown oil which was distilled in vacuo (bp $_{0.7}$ 210-212°C) to obtain 24.6 g (71.2%) of (7). The distilled (7) crystallized upon cooling. mp 41-42°C. Analysis: Calcd.: C 70.46; H 4.77; Found: C 70.50; H 3.93. IR (neat; cm^{-1}); 1750 (CO), 1480, 1400, 1220, 1150. nmr($CDCl_3$; ppm): 7.1 (m, 8H, aryl), 6.0 (m, 3H, vinyl, ABC pattern), 3.8 (s, 2H, benzyl).

2-(o-Benzyl-p-chlorophenoxy)ethanol. 2-Chloroethanol (11.4 g, 141.6 mmol) was added to a solution of o-benzyl-p-chlorophenol (20.2 g, 91.5 mmol) in 9.9% aqueous NaOH (52 ml, 5.1 g NaOH, 127.5 mmol). The reaction mixture was refluxed for 15 hr., cooled and extracted with diethyl ehter. The extract was dried over anhydrous $MgSO_4$ and the ether was removed in vacuo to obtain a brown oil. The oil was dissolved in boiling hexane and white crystals formed upon cooling. Yield: 11.3 g (43.3%). Analysis: Calcd.: C 68.57, H 5.71; Found: C 68.32, H 5.81. IR (KBr; cm^{-1}); 3300 (OH), 1600 (C=C), 1480, 1250 (ArOC), 1100 (COC), 1060 (ArOC). nmr ($CDCl_3$; ppm): 7.1 (s, 6H, aryl), 6.5 - 6.7 (m, 2H, aryl), 3.8 (m, 7H, O$\underline{\text{H}}$), ethylenic and benzylic hydrogens). Upon addition of D_2O to the nmr sample the integration became 8:6 8:6 aryl/ethlenic and benzylic.

2-(o-Benzyl-p-chlorophenoxy)ethyl Acrylate, (8). 2-(o-Benzyl-p-chlorophenoxy)ethanol (20.4 g, 77.7 mmol) and triethylamine (8.6 g, 85.5 mmol) were dissolved in benzene (100 ml) and cooled to 5°C. Acryloyl chloride (7.5 g, 85.5 mmol) was added dropwise. Triethylamine hydrochloride precipitated immediately upon addition of the acryloyl chloride. The mixture was warmed to 20°C for 4 hrs. and then washed once with 5% aqueous $NaHCO_3$ (100 ml) and twice with water (100 ml). The organic layer was dried with anhydrous $MgSO_4$ and the benzene was removed in vacuo to give a yellow oil which crystallized upon cooling. Recrystallization from acetone yielded 7.3 g (29.7%) of (8). mp 74-75°C. Analysis: Calcd.: C 68.25; H 5.37; Found: C 67.88; H 4.93. IR (KBr; cm^{-1}); 1720 (C=O), 1640 (C=C), 1250, 1070, 1050, 990. nmr ($CDCl_3$; ppm): 7.2-7.6 (m, 6H, aryl), 6.8-5.8 (m, 5H, aryl and vinyl), 4.5 (m, 2H, OCH_2), 4.1 (m, 2H, $BrCH_2$), 3.9 (S, 2H, benzylic). The ethylenic hydrogens at 4.1 and 4.5 showed an A_2B_2 pattern.

2-Pentachlorophenoxyethyl Bromide. A solution of pentachlorophenol (50. 1 g, 188.7 mmol) in ethanol (50 ml) was added to 10% aqueous NaOH (85.5 ml) and the resulting mixture was added slowly to refluxing 1,2-dibromoethane (176.6 g, 939.4 mmol). After 12 hrs. water (100 ml) was added to the reaction and the organic layer separated and dried over anhydrous $MgSO_4$ for 8 hrs. Removal of excess 1,2-dibromoethane in vacuo yielded a light brown solid. Recrystallization from 95% ethanol gave 55.5 g (71.9%) mp 84-85°C (lit. 81-82°). Analysis: Calcd.: C 25.70; H 1.34; Found: C 25.9, H 1.06. IR (KBr; cm^{-1}): 1450, 1400, 1300, 1220 (ArOC), 1140 (COC), 1060 (ArOC). nmr ($CDCl_3$; ppm): 4.3 (t, 2H, OCH_2), 3.7 (t, 2H, $BrCH_2$); J = 6 Hz.

Vinyl Pentachlorophenyl Ether, (9). Method A: 2-Pentachlorophenoxyethyl bromide (0.4 g, 1.1 mmol) was dissolved in tetrahydrofuran (3 ml) and 1,5-diazabicyclo-5,4,0-undec-5-ene (DBU) (0.16 g, 1.1 mmol) was added dropwise to the solution. The mixture was refluxed for 30 minutes and water (100 ml) was added, and the solution was extracted with chloroform. The chloroform extract was dried with anhydrous $MgSO_4$ and the chloroform removed in vacuo to yield 0.2 g (64.5%) of (9).

Method B: Pentachlorophenol (9.7 g, 36.5 mmol) was dissolved in vinyl acetate (18.8 g, 218.6 mmol) and the solution was cooled to -30°C. Mercuric acetate (0.9 g, 2.8 mmol) and 2 drops of conc. H_2SO_4 were added to the solution. Reaction was allowed to proceed for 3 hrs. at -30° and then for 12 hrs. at -5°C. The excess vinyl acetate was removed by passing air over the mixture and the solids were chromatographed on grade III alumina using a 4:1 hexane/benzene eluent. The material eluting first was recrystallized from chloroform to obtain 2.7 g (25.0%) of (9).

mp 84-85°C (lit. 82-84°C).[9] Analysis: Calcd.: C 32.82; H 1.03; Found: C 33.15; H 1.00. IR (KBr; cm^{-1}): 1620 (C=C), 1390, 1150, 1120, 980, 950, 910, 850. nmr (CDCl$_3$-DMSO-d$_6$; ppm): 6.5 (Q; 1 J_{AX} = 6 Hx; J_{BX} = 3 Hz), 4.3 (quartet of doublets, 2H, J_{AB} = 2 Hz).

Method C: Pentachlorophenol (5.1 g, 18.8 mmol), 1,2-vinyl-bis(triphenylphosphonium)dibromide (13.3 g, 18.8 mmol) and triethylamine (2.1 g, 20.9 mmol) were dissolved in chloroform (50 ml) and refluxed for 15 hrs. The chloroform solution was washed with 0.2 N HBr (50 ml) and then with water (50 ml) three times, dried over anhydrous MgSO$_4$ and filtered. The chloroform was removed in vacuo to leave a dark brown oil. The oil was relfuxed in a 60:40 benzene/5N aqueous NaOH mixture (100 ml) for 5 hrs., the benzene layer separated and the NaOH solution was extracted with benzene. The combined benzene layers were dried with anhydrous Na$_2$SO$_4$. Removal of the benzene in vacuo yielded a dark brown oil which was chromatographed on grade III alumina using a 4:1 hexane/benzene eluent to give 0.8 g (14.5%) of (9).

2-(2-Bromoethoxy)-3,4',5-Tribromosalicylanilide. The sodium salt of 3,4',5-tribromosalicylanilide (8.5 g 18.0 mmol) was dissolved in a 75:25 dimethylformamide/water mixture (40 ml) and the solution was added dropwise to 1,2-dibromoethane (45.8 g, 243.5 mmol). The mixture was heated to 90°C for 30 min., washed with water (100 ml) three times, and the organic layer was dried with anhydrous MgSO$_4$. Upon removal of the 1,2-dibromoethane in vacuo, a brown solid was obtained which recrystallized from an acetone - ethanol mixture to give 4.7 g (46.9%) of 2-(-bromoethoxy)-3,4',5-tribromosalicylanilide. mp 173-174.5°C. Analysis: Calcd.: C 32.52; H 1.98; N 2.51; Found: C 32.52; H 1.78 N 2.39. IR (KBr; cm^{-1}); 3300 (NH), 1670 and 1600 (C=O), 1550, 1270 (ArOC), 1090 (COC), 1020 (COC). nmr (DMSO-d$_6$, ppm): 9.2 (broad s, NH; 1), 8.1 (d, acid ring, 1), 7.8 (d, acid ring, 1), 7.5 (d, amide ring, 4H), 4.3 (t, OCH$_2$, 2), 3.7 (5, BrCH$_2$, 2).

Vinyl 3,4',5-Tribromosalicylanilide Ether, (10). Method A: 2-(2-bromoethoxy)-3,4',5-tribromosalicylanilide (6.9 g, 12.4 mmol) was dissolved in tetrahydrofuran (30 ml) and the solution heated to 40°C. DBU (2.0 g, 13.3 mmol) was added dropwise and a white precipitate began to form. After 20 min., the mixture was cooled to 20°C and stirred for 13 hrs. Water (50 ml) was added and the reaction was filtered. The solids from the reaction was filtered. The solids from the reaction were dried in vacuo and recrystallized from an acetone-ethanol mixture to give 4.0 g (67.8%) of (10), mp 203-204°C. Analysis: Calcd.: C 37.82; H 2.10; N 2.94; Found: C 37.61; H 1.92; N 2.87. IR (KBr; cm^{-1}): 300 (NH), 1650 (C=O), 1580, 1270, 1250, 1230, 1160, 1040. nmr (DMSO-d$_6$; ppm): 8.0 (s, 1H), 7.5 (m, 1H), 7.2 (m, 2H), 4.5 (m, 1H), 3.9 (m, 2H).

Method B: 2-(2-Bromoethoxy)-3,4,5-tribromosalicylanilide (3.3 g, 5.93 mmol) was dissolved in dimethylformamide (10 ml) and added dropwise to a solution of NaOH (0.8 g, 20 mmol) in 95% ethanol (20 ml). After 10 min., a yellow precipitate formed. After 1 hr., the mixture was filtered and extracted with ether. The extract was dried with anhydrous $MgSO_4$ and the ether removed in vacuo to obtain 1.3 g of 2-(-bromoethyoxy)-3,4,5-tribromosalicylanide. The yellow precipitant was recrystallized from an acetone-ethanol mixture to give 0.5 g (17.7%) of (10). The water/ethanol dimethyformaide mixture was acidified and a white precipitate formed. The precipitate was filtered out and dried to obtain 0.8 g of 3,4,5-tribromosalicylanilide.

2-(O-Benzyl-p-chlorophenoxy)ethyl Bromide. A solution of o-benzyl-p-chlorophenol (25.0 g, 114.4 mmol) in 10% aqueous KOH (loo ml) was added dropwise to refluxing 1,2-dibromoethane (54.0 g, 287.2 mmol). After 15 hrs. the organic layer was separated from the reaction, washed three times with 5% aqueous KOH, three times with water, and dried with anhydrous $MgSO_4$. Removal of the excess dibromoethane in vacuo gave a brown oil which recrystallized from hexane to give 27.1 g(73.0%) of the title compound. Analysis: Calcd.: C 55.30; H 4.30; Found: C 55.34; H 4.24. ir (KBr; cm^{-1}): 1600 (C=C), 1500, 1250 (ArOC), 1070 (COC), 1050 (ArOC). nmr ($CDCl_3$; ppm): 7.1 (s, 7H, aryl), 6.4 (m, 1H, aryl), 4.0 (t, 2H, OCH_2), 3.8 (t, 2H, $BrCH_2$), 3.3 (s, 2H, benzyl).

Vinyl o-Benzyl-p-chlorophenyl Ether, (11). Method A: o-Benzyl-p-chlorophenol (5.0 g, 22.9 mmol) and 1,2-vinyl bis-(triphenylphosphonium)dibromide (16.3 g, 23.0 mmol) were mixed and dried in vacuo. The mixture was dissolved in chloroform (50 ml) and triethylamine (2.6 g, 25.7 mmol) in chloroform (8 ml) was added dropwise at 20°C. After complete addition of the triethylamine, the mixture was refluxed for 16 hrs. The solution was washed twice with 0.2 $\underline{N}$ HBr, three times with water and poured into ether to give a brown oil which was isolated and refluxed for 5 hrs. with a 60:40 benzene/$5\underline{N}$ aqueous NaOH mixture (100 ml). The aqueous layer was separated and extracted with benzene. The extract was combined with the benzene from the mixture, washed twice with water, and dried over anhydrous $MgSO_4$. Removal of the benzene in vacuo gave a brown oil which was chromatographed on grade III alumina using a 4:1 hexane/benzene eluent. Removal of the eluent gave 3.9 g (69.6%) of (11). bp 109-112°C. Analysis: Calcd.: C 73.62; H 6.13; Found C 73.60; H 5.13. IR (neat, cm^{-1}): 1640 (vinyl C=C), 1480, 1250 (vinyl COC), 1150, 1030 (COC), 750, 700. nmr ($CDCl_3$; ppm): 6.9 (m, 8H, aryl), 6.3 (q, 1H, vinyl), 4.3 (m, 2H, vinyl CH_2), 3.7 (s, 2H, benzyl). The quartet at 6.3 and multiplet at 4.3 show an ABX pattern: J_{AB} = 1.5 hz, J_{BX} = 6.2 hz, J_{AX} = 14.0 hz.

Method B: o-benzyl-p-chlorophenol (1.8 g, 8.6 mmol) was dissolved in vinyl acetate (4.4 g, 51.4 mmol) and cooled to -20°C. Mercuric acetate (0.6 g, 1.8 mmol) and 2 drops of conc. H_2SO_4 were added. After 12 hrs., the mixture was chromatographed on a silica gel column using a 2:1 hexane/benzene eluent. The first material eluting off the column was 1,1-bis-(o-benzyl-p-chlorophenoxy)ethane (the acetal produced from vinyl o-benzyl-p- chlorophenylether and o-benzyl-p-chlorophenol). None of the vinyl ether was found however. nmr of the acetal in $CDCl_3$ showed a multiplet at .9 (aryl), a quartet at 5.7 (CH), a singlet at 3.8 (benzylic) and a doublet at 1.3 (CH_3).

Method C: 2-(o-benzyl-p-chlorophenoxy)ethyl bromide (3.3 g, 10.1 mmol) was dissolved in tetrahydrofurna (50 ml) and the solution heated to refluxing. Potassium t-butoxide (1.4 g, 12.1 mmol) in tetrahydrofuran (8 ml) was added slowly to the refluxing solution. After 16 hrs., the mixture was filtered and water (50 ml) was added. The mixture was extracted with ether, the extract dried with anhydrous $MgSO_4$, and the ether removed in vacuo to obtain a light yellow liquid which was distilled in vacuo to get 1.7 g (67.8%) of (11). $b.p._{0.04}$ 107-108°C.

General Procedure for Synthesis of Stable Lactices. Illustrated for Latex 1 (Table II).

A. Preparation of Kettle Charge. Aerosol 22 (tetrasodium N-(1,2-dicarboxy-ethyl)-N-octadecylsulfonsuccinamate (10.8 g) methanol (9.5 g), sodium bicarbonate (1.8 g) were dissolved in water (176 g) and an ammonium persulfate solution (38 g of a 10% by wt. aqueous solution) was added. The solution was purged with nitrogen for 15 min. and charged to the polymerization vessel and heated to 60°C. B. Preparation of Pre-Emulsified Monomer Charge. Aerosol A-102 (disodium ethoxylated alcohol half ester of sulfo succinic acid) (10.8 g), methanol (9.5 g) and sodium metabisulfite (0.8 g) were dissolved in water (59.1 g). The monomers were slowly added with stirring to this solution and the resulting emulsion was purged with nitrogen for 15 min. The amounts were: monomer 1 (20 g, 62 mmol.), MMA (200 g, 2.2 moles) and nBA (160 g, 1.25 moles). The resulting monomer emulsion was charged to the reaction vessel (A) from a dropping funnel fitted with a stirrer and agitation was maintained in the dropping funnel during the entire addition procedure. C. Preparation of Delayed Charge. A delayed charge of itaconic acid (3.8 g), diammonium phosphate (0.6 g) in water (19.1 g) was purged with nitrogen for 15 min. D. Polymerization. After the reaction kettle (with its initial charge) reached 60°, the preemulsified monomer charge was added at a rate of 1-2% per minute. When approximately 15% of the monomer charged had been added, the addition was stopped until polymerization was indicated by an exotherm. The delayed charge was then added to the kettle and the monomer charge was started again. After all the monomer had been added, the polymerization was continued for 1 hr. at 60°, cooled and filtered through a 100 mesh nylon screen to remove any coagulated material. A portion of this latex was poured into a 7:1 methanol/saturated aqueous NaCl solution to precipitate the polymer which then was twice reprecipitated from THF into methanol. Analysis: C 61.09, H 8.14, Cl 2.95.

Acknowledgments

The Paint Research Institute provided partial support of this work through its Mildew Consortium Program.

Literature Cited

1. R. R. Myers, J. Paint Technol., 43, 47 (1971).
2. L. E. Ludwig, J. Paint Technol., 45, 31(1974).
3. G. A. Stahl and C. U. Pittman, Jr. J. Coating Technol., 50, 62 (1978).
4. C. U. Pittman, Jr., J. Coatings Technol., 48, 31 (1976).
5. J. A. Montemarano and E. J. Dyckman, J. Coatings Technol., 47, 59 (1975).
6. B. G. Brand and H. T. Kemp, "Mildew Defacement of Organic Coatings", Federation of Societies for Coating Technology, Philadelphia, PA (1978).
7. C. U. Pittman, Jr., G. A. Stahl, and H. Winters, J. Coatings Technol., 50, 49 (1978).
8. R. L. Adelman, J. Amer. Chem. Soc., 75, 2678 (1953).
9. M. Julia, Bull Soc. Chim. France 185, 1956.

RECEIVED August 12, 1981.

Poly(thiosemicarbazide) Copper(II) Complexes as Potential Algicides and Molluscicides

L. G. DONARUMA

Polytechnic Institute of New York, Brooklyn, NY 11201

S. KITOH

The Lion Company, Ltd., Odawara-Shi, Kanagawa-Ken, Japan

J. V. DEPINTO, J. K. EDZWALD, and M. J. MASLYN

Clarkson College of Technology, Department of Civil and Environmental Engineering, Potsdam, NY 13676

> Polymeric poly (thiosemicarbazide) copper (II) complexes have been made and are being evaluated as algicides and molluscicides. This endeavor showed considerable promise in the laboratory and slow release data for copper (II) indicate that reusable cartridge type copper (II) systems for schistosomaisis and algae control may be feasible.

A number of poly (thiosemicarbazides) have been prepared which complex copper (II) so strongly that mineral acids and even ethylenediamine tetraacetic acid do not effectively remove the copper ion from the polymers (see Scheme 1)[1]. However, in water, initial studies indicated that over longer periods of time, copper (II) ions were released in such fashion that the polymers might be useful for the slow release of copper (II) to provide algae or schistosomiasis control agents (see Table 1) in natural waters. The effect of algae polluting natural waters is a common sight which we all are familiar with.

Schistosomiasis is a serious, often fatal, disease which attacks humans where aquatic snails are the disease carriers[2]. Such snails live in irrigation ditches, streams, swamps, and other natural waters in the warmer climates. The disease carrying snails can be killed by adding copper (II) to their aquatic habitat, and the classic source of the needed copper (II) is cupric sulfate. However, in this case and for algae the salt rapidly migrates away in moving waters or is exchanged into the surrounding soil being very rapidly lost. Thus, frequent additions are required to provide effectiveness. One way tried to retard loss of copper (II) has been to mill copper sulfate into rubber pellets from which water slowly leaches out copper (II) to provide the needed ions to kill snails. However, in natural waters these pellets float away or sink into the channel soil. Of course, any ion exchange resin could provide a substrate to hold copper (II) so that natural water soluble salts would ex-

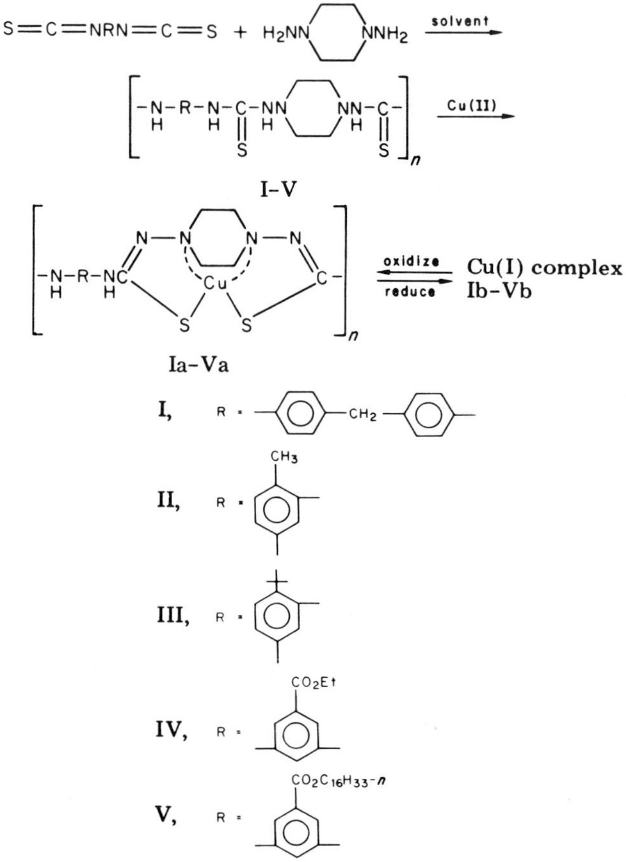

Scheme I. Preparation of poly(thiosemicarbazides).

Table 1
Release of Copper (II) From Polymers I-V as a Function of Time
% Release Over Time*

Polymer	1 da.	2 da.	3 da.	4 da.	7 da.
I	14.6(11.9)	15.6(20.4)	17.4(21.8)	18.8(23.2)	17.3(27.4)
II	4.3(5.8)	5.8(8.6)	7.2(11.6)	8.7(14.4)	10.1(17.3)
III	11.6(18.9)	14.5(20.3)	14.5(21.7)	14.4(24.6)	14.4(21.4)
IV	11.6(13.1)	13.1(14.5)	11.6(15.9)	13.0(17.4)	11.5(18.9)
V	<0.1(0.6)	<0.1(0.6)	0.3(0.6)	0.3(0.6)	<0.1(.84)

Polymer	9 da.	11 da.	14 da.		
I	21.6(28.8)	23.3(31.1)	23.2(33.5)		
II	10.1(20.2)	9.8(25.6)	12.9(25.8)		
III	19.4(28.8)	19.4(32.3)	24.5(34.8)		
IV	11.5(20.2)	19.4(19.4)	18.1(23.2)		
V	<0.1(1.1)	<0.1(4.9)	0.5(5.0)		

* Figure in parentheses is with agitation of the samples.

change away the copper (II) while the regenerable ion exchanger was in a fixed accessible site. However, the copper must be slowly released and not exchanged away from the ion exchanger too quickly. Thus an ion exchanger which binds copper (II) tightly is needed. Yet, the binding of the copper (II) must not be so tight that it will not exchange away from the polymer. The desired rate of copper (II) release should be 1-2% per day.[4] Knowing that copper (II) was tightly bound to I-V, the rate of release from the polymer in water was observed over a period of time.

As can be seen from Table 1 data, the preliminary data indicate that a slow release of copper (II) may be occurring over a fourteen day period. This release apparently is characterized by an initial surge followed thereafter by release of smaller amounts as time goes on. Over a fourteen day period after the initial surge, the release for polymers I, III, and IV averages about in the neighborhood of 1% per day.

It was desired to study the release of copper (II) ions from I-V in detail particularly in synthetic natural waters which contain dissolved ligands capable of eluting copper ions from the polymer and/or precipitating free copper (II).

Experimental

Copper release experiments were performed by placing a weighed amount of resin in a 250 ml Erlenmeyer flask to which a known amount of water of a given chemical composition had already been added. If pH adjustment was required, it was done immediately after the polymer was added. The flasks were then sealed on a rotary shaker (Lab-Line) at 125 rpm. After a given time either the total or dissolved copper concentration in the water was determined.

A semidynamic system was used to examine the resin release properties if introduced to natural water. Resin was placed in nylon mesh bags and put in a flask as previously described. After a certain time interval the nylon bags were removed and placed in a fresh flask of water. This was repeated until no further release of copper was observed.

The resin for these tests was loaded using a batch method. A known weight of resin was placed in a flask containing a known volume and copper concentration at pH = 3. The flasks were then placed on the rotary shaker for 48 hours. They were then removed and the copper solution was filtered through a 0.45 µm membrane filter. The solution was subsequently analyzed for copper concentration. The polymers were then stored in a moist state until used. If rinsing was performed, it was done immediately prior to resin addition to the release water.

Computer equilibrium calculations were performed using MINEQL[5] and an IBM 360-65. The solubility of copper in equilibrium with $Cu(OH)_2$(s) and $Cu_2(OH)_2CO_3$(s) (malachite) was examined.

Composition of Synthetic
Natural Water[3]

pH_{+2} 7.80 - 8.15
Ca^{+2} = 67 mg/L as $CaCO_3$
Mg^{+2} = 67 mg/L as $CaCO_3$
Alk = 100 mg/L as $CaCO_3$
Fe^{+3} = 0.1 mg/L
$SO_4^=$ = 54 mg/L
Cl^- = 52 mg/L

Copper Release Experiments

Both I and IV were initially used in the release experiments. First, the effects of pH were examined. As shown in Figure 1 A and B, at pH = 3 a quick initial release (approximately 15 minutes for IV, 6 hours for I) occurred and after this time no more copper was released. The difference in Figure 1 A and B is that Figure 1 B takes into account resin weight present in the flask.

IV appears to reach the equilibrium release concentration earlier. This result is attributed to the following two reasons. First, under acid conditions IV may be partially eluted with acid solutions. Second, the smaller particles would have a greater surface area and tend to have higher rates of transfer. All the measurements were of dissolved copper unless noted otherwise.

Since natural waters are obviously not at pH=3, the polymers were next placed in a $2 \times 10^{-3}M$ sodium bicarbonate solution at a pH=6.0 - 6.5. As can be seen in Figure 2, the same type of quick initial release occurred. The measured copper values are somewhat lower due to precipitation of copper but probably also due to the fact that less copper is released at higher pH values. The solutions were not buffered well at this pH and bicarbonate condition, and as a result the pH of most solutions when removed for measurements was around 4.0.

The result for the same type of solution at an initial pH of 8 were as expected. Once the copper loaded resin was added, copper was immediately precipitated. The pH of the solution decreased to 7.5-7.6. IV once again tended to release a fraction of its copper quickly, while I released its fraction slightly slower as evidenced by the plateaus in Figure 3 A and B.

As a result of these tests I in the pellet form was chosen for additional investigation because its release characteristics were slower. The rest of the release experiments were performed in a synthetic water to better model natural systems. The first experiment involved placing the pellets in the synthetic natural water (SNW). As was expected results similar to the pH = 8 experiment were observed as shown in Figure 4.

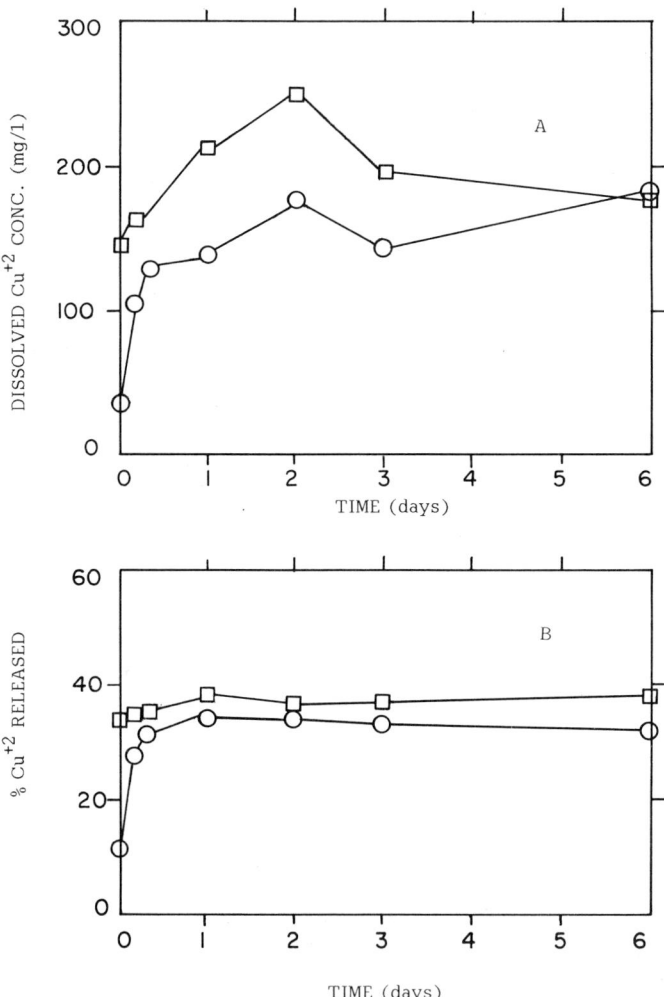

Figure 1. Release of copper by I (○) and IV (□) in deionized water at pH 3. Key: A, dissolved Cu^{2+} concentration (mg/L); and B, percent Cu^{2+} released.

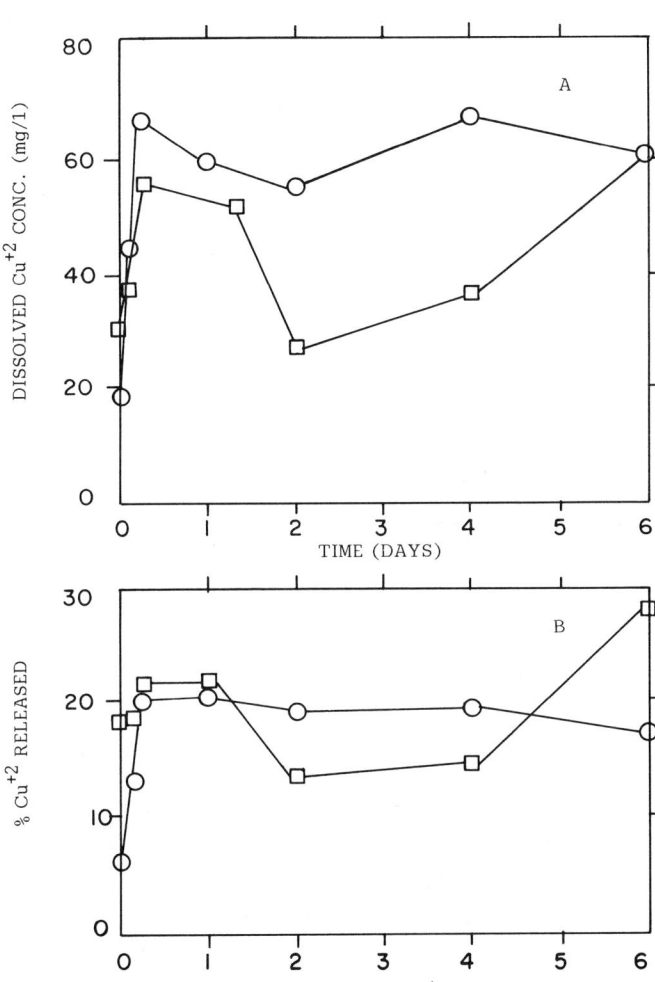

Figure 2. Release of copper by I (○) and IV (□) in 2×10^{-3} M $NaHCO_3$ solution at pH 6–6.5. Key: A, dissolved Cu^{2+} concentration (mg/L); and B, percent Cu^{2+} released.

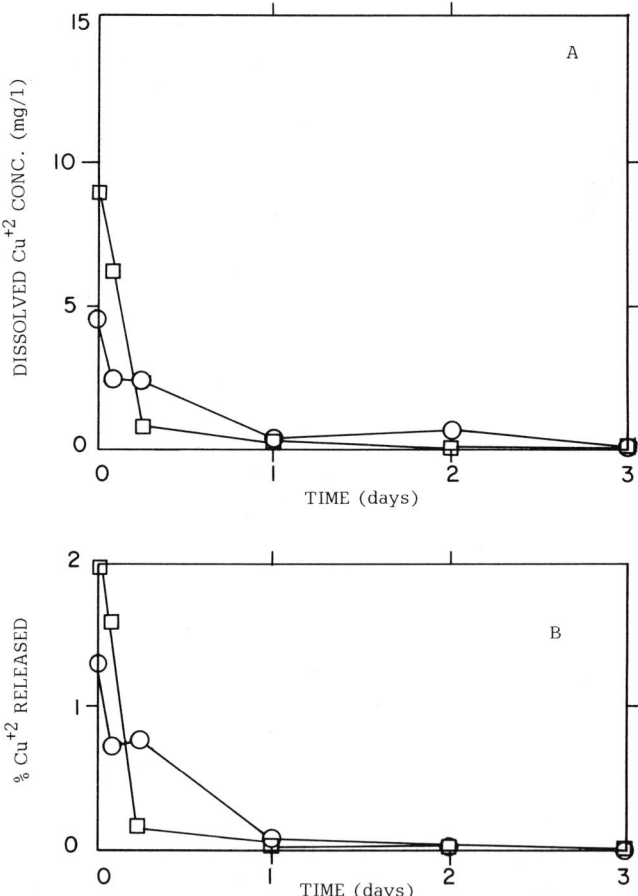

Figure 3. Release of copper by I (○) and IV (□) in 2×10^{-3}M $NaCO_3$ solution at pH 8. Key: A, dissolved Cu^{2+} concentration (mg/L); and B, percent Cu^{2+} released.

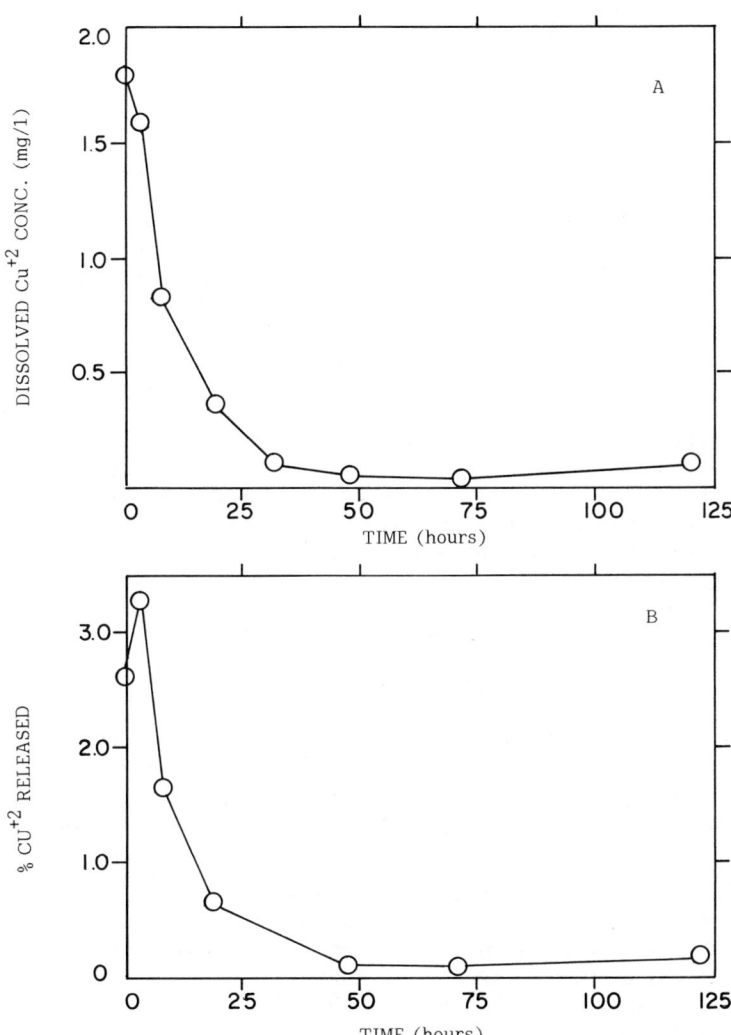

Figure 4. Release of copper by I in synthetic natural water at initial pH 7.8. Key: A, dissolved Cu^{2+} concentration (mg/L); and B, percent Cu^{2+} released.

In order to more closely simulate the dilution effects of a natural water body, a semidynamic system was evaluated using the SNW. As is evidenced by Figure 5A, in both the rinsed and unrinsed cases the pellets released approximately 80 percent of their total copper in the first 15 minutes. After 30 minutes very little additional copper was released from the resin.

The effect of particle size was also examined in the semidynamic system. I (20 x 50 mesh) was exposed under the same conditions (Figure 5 B) and the rinsed cases behave much the same; therefore, it is safe to assume that a good portion of the copper in most of these experiments came from the surface water associated with the moist resin, with more surface water being associated with the smaller size particles. This copper most likely originated from the solutions used to load the polymers.

A final series of experiments were run to evaluate the effects of the presence of an organic complexing agent on release properties. EDTA was chosen as a model organic ligand. As in evidence by Figure 6 the EDTA complexes as much copper as it can. The dissolved copper concentrations at the end of the test for the 10^{-5} M EDTA case are very close to the theoretical values calculated 0.63 mg/l Cu^{2+} as will be presented below. When the EDTA concentration exceeded the stoichiometric amount of copper initially on the resin, a total release of 80 percent of the initially bound copper was observed. This result is not unexpected because of the high formation constant for copper - EDTA complexes. In addition some of the copper in solution was probably due once again to the surface water.

Computer Equilibrium Calculations

To better understand the mechanisms involved in the release studies, computer equilibrium calculations were performed using Mineql (4). The equilibrium speciation of copper with hydroxide as a function of pH was first examined. As can be seen in Figure 7 for a 10^{-3} Cu_T solution; the solid phase takes over from the free copper ion as the dominant phase at a pH of 6.0. For the much lower copper concentrations typically found in water the transition pH is shifted higher.

Next, calculations were performed to demonstrate the possible speciation occurring in the synthetic natural water used. In Table 2 a summary of the values calculated can be found, It can easily be seen that $Cu_2(OH)_2CO_3(s)$ (malachite) is the dominent species for this particular pH and ion composition, and accounts for essentially all the copper. The free copper concentration is relatively low 3.8 X 10^{-9}M (0.24 µg/L). The total soluble Cu(II) concentration is very close to the value found in actual experiments (see Figure 4 and Table 2). The $Cu(OH)_2(s)$ concentration is not available due to the method of solution used in Mineql (4). It performs calculations to determine the dominant phase and once it selects one the other is ignored.

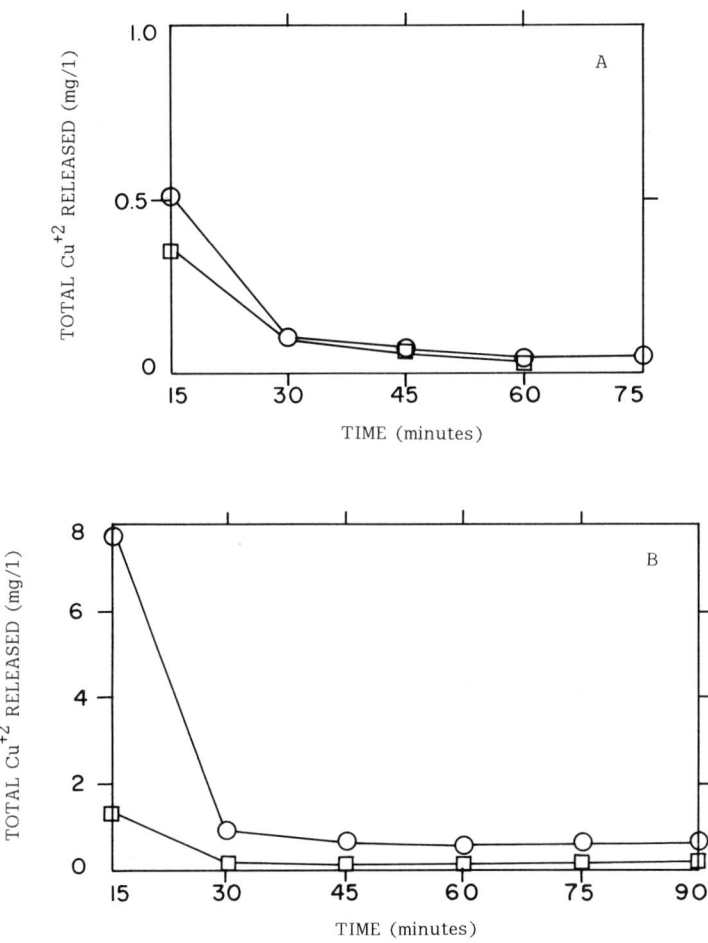

Figure 5. Release of copper by I (20 × 50-mesh pellets) in a semidynamic system using synthetic natural water. Key: ○, unrinsed; □, rinsed; A and B, total Cu^{2+} released (mg/L).

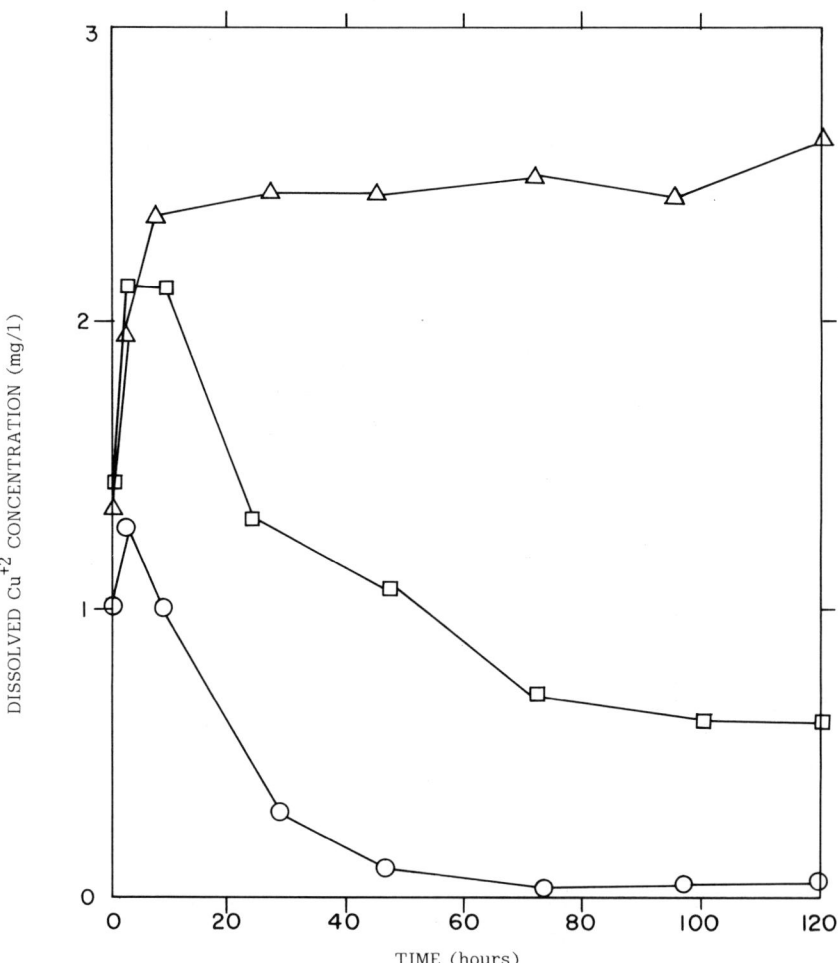

Figure 6. Effects of EDTA in synthetic natural water on the release of I pellets at average initial pH 8.05. Key: ○, 10^{-6}M EDTA; □, 10^{-5}M EDTA; and △, 10^{-4}M EDTA.

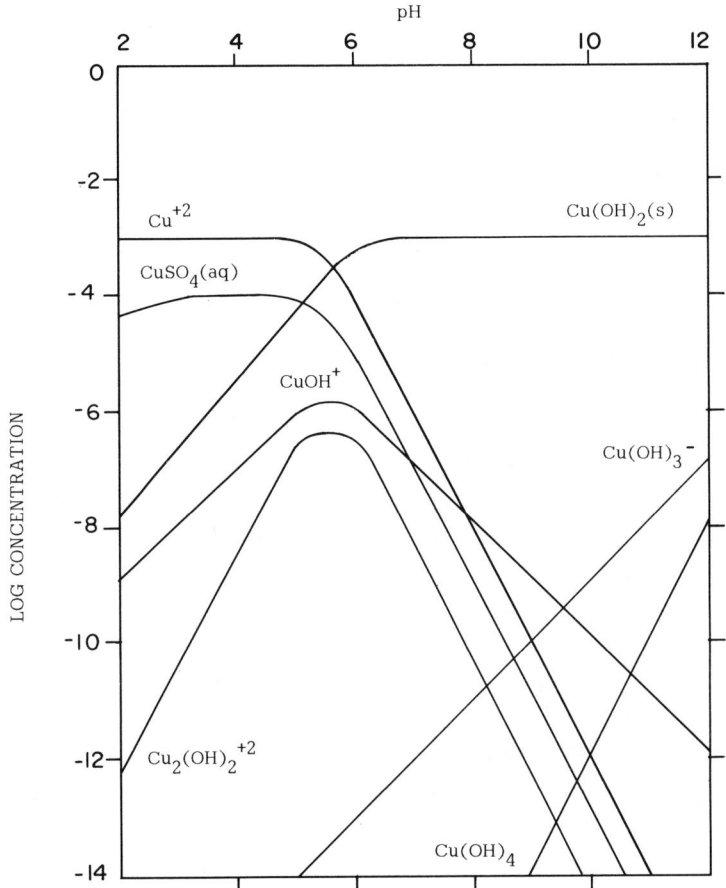

Figure 7. Copper hydroxide equilibrium for a 10^{-3}M solution of $CuSO_4$ as a function of pH.

Table 2 Speciation of Copper in Synthetic Water at pH = 8 ($Cu_T^{+2} = 2 \times 10^{-5}$M; $C_T = 2 \times 10^{-3}$M)

Species	-Log Concentration
Cu^{+2}	8.41
$CuCO_3(aq)$	6.80
$Cu(CO_3)_2^=$	8.36
$CuSO_4$	9.97
$CuCL^+$	10.84
$CuCL_2$	13.98
$CuOH^+$	8.40
$Cu_2(OH)_2$	11.21
$Cu_2OH_2CO_3(s)$	5.00
$Cu(OH)_2 (s)$	N.A.

N.A. - Exact value not available due to the method of computer solution. $Cu_2(OH)_2CO_3(s)$ was the only solid phase considered.

The shift of dominance in solid phase for the synthetic natural water used, occurs between total inorganic carbon concentrations 10^{-4} and 10^{-3} M as shown in Table 3. Therefore, for most of the experiments conducted in the laboratory it is assumed that the solid phase present was malachite. A natural water could easily reduce the free copper concentration to a very low value even in the presence of high total copper. This can easily be seen in Tables 3 and 4.

The presence of organic complexing agents and a variation of pH were also examined to see what effects these might have in synthetic water. EDTA was once again chosen for the model organic ligand and pH was varied from 6 to 9. For lower pH values higher free copper concentrations were found as shown in Figure 8. As EDTA concentration increased the copper that it complexed seemed to originate from the solid phase, thus not affecting the free Cu^{2+} until the solid phase was completely dissolved. At this point the free copper along with some other species such as $CuOH^+$ and $Cu_2(OH)_2^{+2}$ provided the source for the EDTA. Figures similar to Figure 8 can be used to predict free copper concentration changes that might result from pH or organic ligand concentrations.

On the basis of these findings and the examination of the literature it is quite evident the overall picture of copper speciation should be examined in toxicity studies. Organic and inorganic complexation play a very important role in the speciation and toxicity and can not be ignored. Before applying these resins to aquatic organism control more testing in dynamic systems along with toxicity studies should be performed. <u>The resins show high promise as controlled release substrates for copper(II)</u>.

Table 3 Comparison of Speciation of Copper in Varying $CO_3^=$ Concentrations at pH = 8 for 1×10^{-3} Cu_T (-Log Concentration)

Species	$CO_3^= = 0$	$CO_3^= = 10^{-4}$	$CO_3^= = 10^{-3}$	$CO_3^= = 10^{-2}$
Cu^{+2}	8.06	7.94	8.13	8.66
$CuOH^+$	7.90	7.90	7.98	8.71
$Cu_2(OH)_2^{+2}$	10.24	10.24	10.63	11.76
$Cu(OH)_3^-$	10.84	10.84	11.03	11.56
$Cu(OH)_4^=$	15.82	15.81	15.97	16.35
$CuSO_4$	8.88	8.89	9.14	9.98
$Cu_2HCO_3^+$	–	9.97	9.49	8.89
$CuCO_3$	–	7.57	7.09	6.49
$Cu(CO_3)_2^=$	–	10.46	9.25	7.22
$Cu(OH)_2(s)$	3.00	3.00	N.A.	N.A.
$Cu_2(OH)_2CO_3(s)$	–	N.A.	3.3	3.3
$CuCO_3(s)$	–	N.A.	N.A.	N.A.
$Cu_3(OH)_2(CO_3)_2(s)$	–	N.A.	N.A.	N.A.

N.A. - Not available due to the method of computer solution.

Table 4. The Effect of pH and EDTA on Cu^{+2}, $CuOH^+$, $Cu_2(OH)_2^{2+}$ Concentration in Synthetic Water with $Cu_T = 2 \times 10^{-5} M$

	pH = 6				pH = 7		
-log EDTA	Cu^{2+}	-log $CuOH^+$	$Cu_2(OH)_2^{2+}$	-log EDTA	Cu^{2+}	-log $CuOH^+$	$Cu_2(OH)_2^{2+}$
10^{-6}	5.21	7.2	8.81	10^{-6}	6.89	7.87	10.16
10^{-5}	5.21	7.2	8.81	10^{-5}	6.89	7.87	10.16
1.5×10^{-5}	5.38	7.36	9.14	1.5×10^{-5}	6.89	7.87	10.16
2.0×10^{-5}	6.30	8.29	10.99	2.0×10^{-5}	7.87	8.86	12.13
5.0×10^{-5}	11.46	13.44	21.30	5.0×10^{-5}	11.46	12.45	19.31
10^{-4}	11.93	13.92	22.25	10^{-4}	11.94	12.93	20.27

	pH = 8				pH = 9		
-log EDTA	Cu^{2+}	-log $CuOH^+$	$Cu_2(OH)_2^{2+}$	-log EDTA	Cu^{2+}	-log $CuOH^+$	$Cu_2(OH)_2^{2+}$
10^{-6}	8.41	8.40	11.21	10^{-6}	9.92	8.90	12.22
10^{-5}	8.41	8.40	11.21	10^{-5}	9.92	8.90	12.22
1.5×10^{-5}	8.41	8.40	11.21	1.5×10^{-5}	9.92	8.90	12.22
2.0×10^{-5}	8.81	8.80	12.01	2.0×10^{-5}	9.92	8.90	12.22
5.0×10^{-5}	11.47	11.46	17.33	5.0×10^{-5}	12.30	11.29	16.99
10^{-4}	11.95	11.93	18.28	10^{-4}	12.74	11.73	17.87

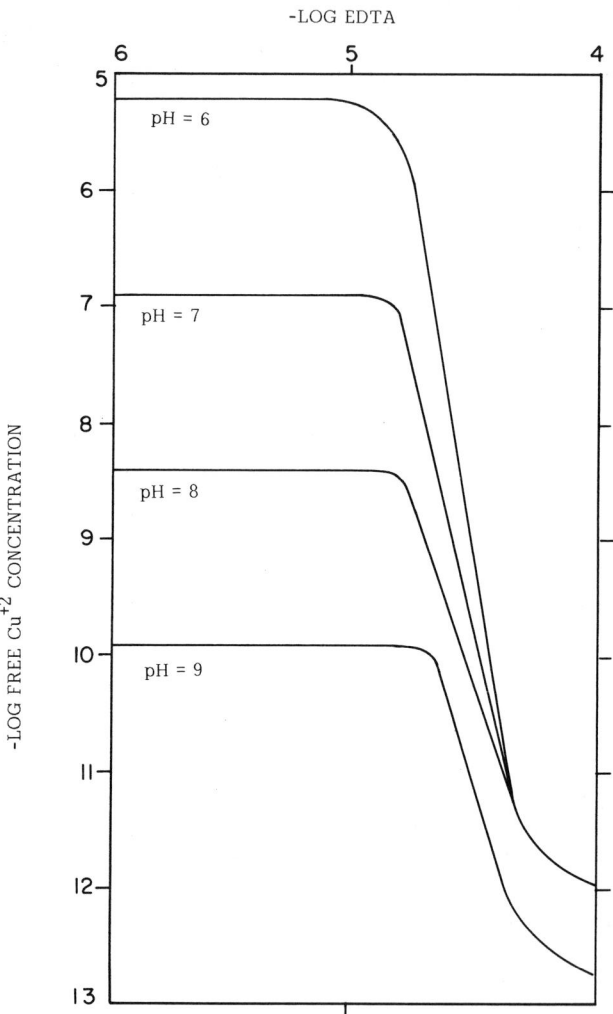

Figure 8. Effect of EDTA and pH on free copper concentration in synthetic water.

Acknowledgment

We are indebted to the International Copper Research Association for their generous support of this work. Parts of this chapter were taken from the theses submitted by Messrs. Kitch and Maslyn in partial fulfillment of the requirements for the Master of Science degree.

Literature Cited

1. Donaruma, L. G., Kitch, S., Walsworth, G., Depinto, J. V., and Edzwald, J. K., Macromolecules, 12, 435 (1979).
2. Cheng, T. C., "Molluscicides in Schistosomiasis Control," Academic Press, New York, NY, 1974.
3. International Copper Research Association, private communication.
4. Westall, J. C., Zachary, J. L., and Marel, F. M., "MINEQL," Department of Civil Engineering, Technical Note No. 18, Massachusetts Institute of Technology, July 1976.

RECEIVED January 11, 1982.

6

Potential Polymeric Herbicides Derived from Poly(vinyl alcohol): Modification of Polymers[1]

CHARLES G. GEBELEIN

Youngstown State University, Department of Chemistry, Youngstown, OH 44555

>   Isocyanates can be reacted with poly(vinyl alcohol) in homogeneous solutions in dimethylsulfoxide to produce modified polymers that contain between 10 and 100% of the hydroxyl groups replaced by the appropriate carbamate group. In this study, modified polymers were prepared using phenyl, 3-chlorophenyl and 4-chlorophenyl isocyanates and these polymers bear a strong structural similarity to known carbamate herbicides.

Many types of biologically active polymeric systems have been described in the literature including drug, insecticide, herbicide and fungicide systems. These systems could involve biologically active polymers or the controlled release of a low molecular weight species from a polymeric matrix and these systems have been reviewed in several recent reports (1-6). In the controlled release system, the biologically active agent is contained within the polymer in some fashion and could be a 'solution' or a dispersion of this agent in a polymeric matrix or the enclosing or encapsulation of the agent in the polymer (as in a microcapsule). The primary objective of such controlled release systems is to permit the release of a fairly constant level of the biologically active agent for a relatively long period of time. A biologically active polymer, on the other hand, would exhibit its biological activity as a polymer and would not require the release of a low molecular weight species for its action. The potent biological activities of the nucleic acids and the enzymes are examples of polymeric biologically active agents and a number of synthetic polymers (e.g. the divinyl ether- maleic anhydride cyclocopolymer) have shown this type of behavior (7). Many polymeric systems lie between these extremes and sometimes these may exhibit biological activity directly (as a polymer) or serve as a source of controlled release of a low molecular weight species that is attached to the polymer. In the latter case, the system would bear a close

[1] This is Part IV in a series.

thermokinetic relationship to an ideal monolithic controlled release system if the bioactive agent were uniformly distributed along the polymeric backbone. In actual practice, it is not always obvious whether a polymeric molecule derives its activity directly or by the release of a low molecular weight species and this can only be deduced experimentally. If a polymeric system showed release of a known bioactive agent, it is probable that this system acts as a controlled release system but this does not preclude the possibility of this system also functioning as a polymeric biological agent. On the other hand, if a system does not show the release of any bioactive agent and still exhibits biological activity, then this system must function as a polymeric biologically active system.

Both the polymeric systems and the controlled release systems could be valuable in medicinal areas or in pest control. In each case, prolonged activity of the bioagent would be expected as well as the maintainance of a low, fairly constant concentration of this agent. A non-degrading, polymeric bioagent could maintain its activity for an indefinite period but could create new environmental challenges. It is important to take this potential problem into consideration in designing a polymeric bioagent since some polymers can survive for a very long time in the natural environment and could create major pollution problems. Nevertheless, a polymeric bioagent has several unique potential advantages over a controlled release agent. For example, the polymeric bioagent could be made water soluble fairly readily (e.g. by copolymerization with a solubilizing monomer) and, in principle, could be directed to a specific target site (e.g. a diseased organ or a specific type of weed) by careful, selective copolymerization reactions. Such specific effects cannot be expected for controlled release systems. These special considerations have been discussed recently for polymeric drugs (7,8).

Polymeric bioagents can be prepared by the modification of existing polymers (9,10) or by the synthesis and polymerization of new monomeric agents (3,7,8). In the present paper, we are describing the synthesis of some new, potential herbicide polymers by the modification of poly(vinyl alcohol) in a continuation of our previous work in this area (11-13).

Experimental

Materials Used. The poly(vinyl alcohol) samples used in this study were two fully hydrolyzed materials which were vacuum dried several days at 100°C and 10 mmHg to remove any residual water. These PVA's had molecular weights of 14,000 and 60,000. Dry, analytical grade dimethylsulfoxide (DMSO) was used as the reaction solvent. The isocyanates were used as supplied and included phenylisocyanate (Eastman), 4-chlorophenylisocyanate (Columbia) and 3-chlorophenylisocyanate (Eastman).

Modified Polymer Preparation. The basic procedure for the preparation of these modified polymers is outlined in Equation (1) and has been described in detail in our earlier publications (11-13). The basic procedure is to dissolve the dried PVA in DMSO and then add, with stirring, a solution of the isocyanate in DMSO. (While the isocyanate could be added directly, this procedure minimizes any problems that could arise by adding a high concentration of the isocyanate in preparing a partially substituted product which could give a highly heterogeneous material.) A small amount of triethylamine is added as a catalyst and the system is stirring several days at room temprature. The modified polymer can be isolated by pouring the system into a large amount of water (preferred) or methanol while stirring rapidly. The precipitate is collected by filtration, washed with non-solvent, air dried and vacuum dried to give the modified polymer. Using this general procedure, modified polymers can be prepared with 10 to 100% substitution of the hydroxyl groups in recovered yields that normally exceed 90%.

In a typical example, 6.6 g. PVA (MW of 14,000) was dissolved in 100 ml. DMSO and a solution of 27.6 g. 3-chlorophenylisocyanate (20% excess) in 100 ml. DMSO was added, with sitrring, followed by 1 ml. triethylamine. After 5 days stirring at room temperature on a magnetic stirrer, the polymer was isolated by pouring into several liters of water. After filtering and drying under vacuum, 26.7 g. modified polymer (90.2% of theory) was isolated which had essentially all the -OH groups substituted by carbamate groups. The isocyanates used in this study included phenyl, 3-chlorophenyl and 4-chlorophenyl isocyanates.

Polymer Properties. Infrared spectra were determined for the modified polymers using a Beckman Acculab 4 Spectrophotometer on films cast on NaCl plates from dimethylacetamide solutions. As the degree of substitution increased, the -OH peak (3300 cm^{-1}) decreased and peaks due to the amide II band (1560 cm^{-1}), -NH and aromatic bands increased.

As the degree of substitution increased, the polymers became more brittle and film formation became more difficult. Polymer solutions, in DMSO or DMAC, became less viscous as the degree of substitution increases even though the polymer molecular weight increases. Below about 20% substitution, these solutions tend to gel in the same manner as unsubstituted PVA1 solutions. (The molecular weight of the repeat unit increases about 450% in PVA1 which is completely substituted with a ClC_6H_4NCO and a polymer with an initial molecular weight of about 14,000 would increase to about 62,800.)

Hydrolysis studies were run by placing a 0.5 g. polymer sample in 50 ml. distilled water and allowing this to stand with occasional stirring. After two months, an aliquot of the supernatent liquid was analyzed spectroscopically in a Cary 14 for the presence of the appropriate aniline derivative. A polymer fully substituted with $4\text{-}ClC_6H_4NCO$ only releases 0.46% of the 4-chloroaniline in this two month period.

Discussion

Poly(vinyl alcohol) readily reacts with a wide variety of isocyanates in DMSO solution (Equation 1) to produce modified polymers that contain 10-100% of the original hydroxyl groups substituted with carbamate groups. The yields of such reactions are essentially quantitative. Similar results have been reported from other research groups (14-16). Under these conditions, poly(vinyl alcohol) has been reacted with 3- and 4-chlorophenylisocyanate (this study), phenylisocyanate (12-14), methyl, ethyl, isopropyl and 1-naphthyl isocyanates (14) and methoxymethylisocyanate (15,16). The reaction of PVAl with phenyl, tolyl and 4-chlorophenyl isocyanates in dimethylacetamide solutions has also been reported (17).

$$\mathrm{-(CH_2CH)_n\!-} \quad \xrightarrow[\mathrm{DMSO,\ Et_3N}]{x\ R\text{-}NCO} \quad \mathrm{-(CH_2CH)_{n-x}(CH_2CH)_x\!-}$$
$$\mathrm{OH} \qquad\qquad\qquad\qquad\qquad \mathrm{OH} \qquad \mathrm{O}$$
$$\qquad\qquad\qquad\qquad\qquad\qquad\qquad\qquad \mathrm{C=O}$$
$$\qquad\qquad\qquad\qquad\qquad\qquad\qquad\qquad \mathrm{NHR}$$

(Equation 1)

The carbamate containing polymers prepared in this study bear a close structural relationship to some relatively common carbamate herbicides. A comparison of these structures is shown in Figure 1. Carbamate herbicides are known to function by inhibiting cell division and are classed as antimitotic agents. These carbamate herbicides are normally applied to the soil rather than the foliage and are best applied as a pre-emergent control material. Although each herbicide has a specific group of plants to which it is either harmless or toxic, the following generalizations can be made. The carbamate herbicides are normally harmless to the following plants: flax, garlic, marigolds, mustard, onions, peas, rape, safflower, spinach, soybeans, sugar beets and sunflower. On the other hand, the carbamate herbicides are usually toxic to the following plants: barley, crabgrass, darnel (ryegrass), fescue, goosegrass, maize, quackgrass (counchgrass) oats, timothy, watergrass, wild oats and wheat. It is well established that the carbamate herbicides degrade in the soil. via microbial action, to produce an aniline derivative as shown in Equation 2. This hydrolysis occurs fairly rapidly and the original herbicide structure does not survive long in the natural environment. Whether the resulting aniline derivative can also be a source of antimitotic activity does not appear to be established but the toxicity of aniline derivatives is well known (18-20). The antimitotic activity is generally believed to occur via the

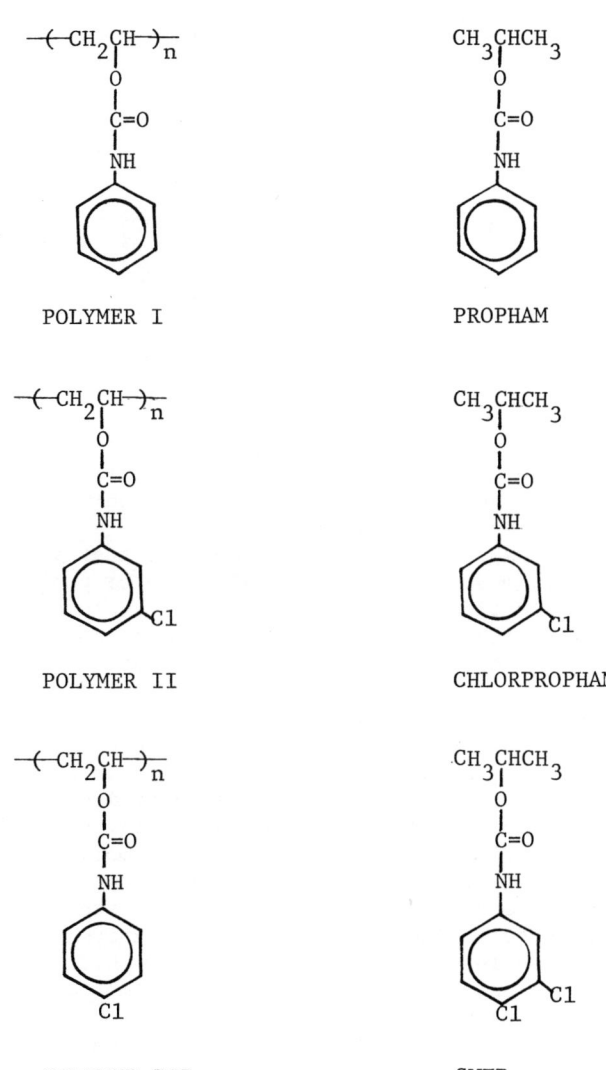

Figure 1. Structural relationships between the modified polymers and some carbamate herbicides.

carbamate molecule as a whole rather than from some degradation product. In addition, not all carbamate herbicides contain a N-phenyl group and an aniline derivative is not always a degradation by-product. (For example, dichlormate is 3,4-dichlorobenzyl methylcarbamate and would yield methyl amine and 3,4-dichlorobenzyl alcohol.) A number of insecticide carbamates exist which are known to function by inhibiting acetylcholinesterase ($\underline{21,22}$). These are usually derivatives of methyl carbamate.

$$\underset{\substack{|\\ \text{O}\\ |\\ \text{C=O}\\ |\\ \text{NH}\\ |\\ \text{C}_6\text{H}_5}}{\text{CH}_3\text{CHCH}_3} \xrightarrow{\text{H}_2\text{O}} \underset{\substack{|\\ \text{OH}}}{\text{CH}_3\text{CHCH}_3} + \text{C}_6\text{H}_5\text{NH}_2 + \text{CO}_2$$

(Equation 2)

The controlled release of various types of herbicides have been studied by many workers and has been summarized elsewhere ($\underline{4-6}$, 23). Less is known, however, about the controlled release of herbicides for land plants than for aquatic plants. Previous studies by other research groups have reported that isocyanate modified polymers, made from chitin, cellulose or poly(vinyl alcohol), hydrolyze slowly in aqueous media to release the corresponding aniline derivative (17,24,25). No data appears to be currently available on how effective these agents are as herbicides.

In the present study, we have prepared a series of polymers which contain the carbamate units shown in Figure 1. While these materials do resemble known herbicides closely, this does not necessarily mean these polymers would also exhibit this type of activity although there are many instances of polymeric drugs that are biologically active in the same way as low molecular weight analogs without these polymeric drugs serving as a controlled release source of this drug ($\underline{1,3,7}$). The systems studied here do release the expected aniline derivative when exposed to water but the rate of release is very slow (approximately 0.23% per month) and these modified polymers can be isolated from the preparation reaction media (DMSO) by pouring this solution into water. The polymers themselves are completely insoluble in water, therefore we would not expect them to be highly mobile under the use conditions that would be encountered in agriculture, etc. This would be an advantage in maintaining the proper herbicide levels and also in preventing unwanted migration of this agent to other areas where it could cause environmental damage. At the present time, biological activity data are not available for these polymers.

The solution viscosity behavior of the new 3- and 4-chlorophenyl isocyanate derivatives are similar to that reported earlier for the phenyl isocyanate derivative in that the viscosity decreases as the extent of substitution, and overall molecular weight, increases (12). This effect is believed to be due to either a decrease in hydrogen bonding interactions as the fraction of hydroxyl groups decreases or to an increase in chain stiffness due to the substitution of the bulky carbamate units for these hydroxyl groups which would reduce chain entanglements and would lower the viscosity of these solutions. Further studies are in progress on this viscosity effect and will be reported in greater detail in a future publication.

Literature Cited

1. Gebelein, C.G.; Koblitz, F.F., Ed.; "Biomedical and Dental Applications of Polymers"; Plenum Press: New York, 1981.
2. Goldberg, E.P.; Nakajima, A., Ed.; "Biomedical Polymers"; Academic Press: New York, 1980.
3. Donaruma, L.G.; Vogl, O., Ed.; "Polymeric Drugs"; Academic Press: New York, 1978.
4. Baker, R., Ed.; "Controlled Release of Bioactive Materials"; Academic Press: New York, 1980.
5. Cardarelli, N. "Controlled Release Pesticide Formulations"; CRC Press: Cleveland, 1976.
6. Paul, D.R.; Harris, F.W., Ed.; "Controlled Release Polymeric Formulations"; Am. Chem. Soc.: Washington, DC, 1976.
7. Gebelein, C.G. Polymer News 1978, 4, 163.
8. Gebelein, C.G.; Morgan, R.M.; Glowacky, R.; Baig, W. in Ref. (1), p. 191.
9. Carraher, C.E.; Tsuda, M., Ed.; "Modifications of Polymers"; Am. Chem. Soc: Washington, DC, 1980.
10. Moore, J.A., Ed.; "Reactions on Polymers"; Reidel Publ: Dordrecht, Holland, 1973.
11. Gebelein, C.G. Organic Coatings & Plastics Chem. 1981, 44, 23.
12. Gebelein, C.G.; Burnfield, K.E. in Ref. (9), p. 84.
13. Gebelein, C.G.; Burnfield, K.E. Organic Coatings & Plastics Chem. 1979, 40, 53.
14. Sastre, R.; Garcia Perez, M.; Acosta, J.L. Rev. Plast. Mod. 1977, 34, 76.
15. Sikorski, R.T.; Hadrowicz, B.; Kokocinski, J.; Kowalczyk, M., Pol. 86,452 (1976); Chem. Abstr. 1977, 86, 191,429e.
16. Sikorski, R.T.; Hadrowicz, B.; Kokocinski, J.; Kowalczyk, M. Pr. Nauk. Inst. Technol. Org. Twerzyro Sztweznych Polstech. Wroclaw 1976, 16, 101; Chem. Abstr. 1976, 83, 44,761s.
17. McCormick, C.L.; Pelezo, J.A.; Anderson, K.W.; Hutchinson, B.H., Jr., "7th International Symposium on Controlled Release of Bioactive Materials"; Ft. Lauderdale, FL, July, 1980, p. 182.
18. Crafts, A.S. "The Chemistry and Mode of Action of Herbicides"; Interscience: New York, 1961, p. 71.
19. Ashton, F.M.; Crafts, A.S. "Mode of Action of Herbicides"; Interscience: New York, 1973, p. 200.

20. Corbett, J.R. "The Biochemical Mode of Action of Pesticides"; Academic Press: New York, 1974, p. 189.
21. <u>Ibid.</u>, p. 107.
22. Kuhr, R.J.; Dorough, H.W. "Carbamate Insecticides: Chemistry, Biochemistry and Toxicology"; CRC Press: Cleveland, 1976.
23. Scher, H.B., Ed.; "Controlled Release Pesticides"' Am. Chem. Soc.: Washington, DC, 1977.
24. Matsunaga, T.; Ikada, Y., in Ref. (9), p. 391.
25. McCormick, C.L.; Lichatowich, D.K.; Pelezo, J.A.; Anderson, K.W., in Ref. (9), p. 371.

RECEIVED August 12, 1981.

DRUG RELATED ACTIVITY: MEDICAL APPLICATIONS

In Vivo Studies on Drug—Polymer Sustained-Release Systems

JAMES M. ANDERSON

Case Western Reserve University, Departments of Pathology and Macromolecular Science, Cleveland, OH 44106

>   The pharmacodynamic performance was studied for three drug-polymer matrix systems and comparisons were made between the in vivo and the in vitro results. The systems studied were: (a) gentamicin in silicone rubber, (b) tetracycline in HEMA/MMA copolymers, and (c) niridazole in silicone rubber. In general, the in vitro models fall short of predicting the actual in vivo results accurately, even in those cases where the controlled release of the agent showed positive, desirable results in the test animals.

Sustained release of biological agents is a concept that has received increased attention over the past decade. The primary goal of sustained drug release is the prolonged delivery of a drug to a particular body compartment or anatomical target site. This goal is accomplished by the application of a therapeutic drug delivery system designed to control both temporal and spatial aspects of drug disposition.

Numerous drug-polymer delivery systems have been proposed with the expressed purpose of releasing a biologically active agent into the surrounding medium at a constant (zero-order) release rate. Attempts to achieve zero-order release rates have included infusion pumps (1), erodable matrices (2), controlled geometries (3), and various membrane-enclosed reservoir devices (4, 5, 6). Unfortunately, many of these systems fall far short of their goal, suffering from prolonged burst effects, device breakdown, and constraints in design, fabrication, or material properties. The result is usually an exponential release pattern with time and a relatively limited period of constant release.

Previous investigations in the field of controlled drug delivery have centered on the design and in vitro testing of the device. Animal experimentation has been largely limited to reports on the temporal pharmacodynamic performance of the drugs released from the implanted, inserted, or surface-applied polymer vehicles. Few studies have attempted to provide details of the kinetic

0097-6156/82/0186-0085$5.00/0
© 1982 American Chemical Society

behavior of these release systems throughout the period of in vivo evaluation. These data are essential to exploring the pharmacokinetic characteristics of such systems and providing for the development of predictive models. These models also increase experimental efficiency by allowing the use of preliminary measurements and analyses to optimize the further design of the controlled drug-release systems, conserving time and animal resources.

We wish to review here our efforts which have been directed toward evaluating the in vivo release behavior of various drug-polymer devices. We have used dogs, rats and mice in these studies for reasons which will be explained. Our studies, which are briefly summarized in Table 1, have had two major themes: the investigation of drug release behavior in a disease model in an animal, and the investigation of the pharmacokinetics of a drug-polymer sustained release system in an animal model. Both of these themes are pertinent to the evaluation of drug-polymer sustained release systems intended for use in humans or animals. We intend only a brief review here and refer the reader to the references for experimental techniques, details and an in-depth discussion of each system.

Table 1. Drug-Polymer Sustained Release Systems

General Theme or Disease Model	Studies	Animal	Drug	Polymer	References
Prosthetic Valve Endocarditis	in vitro in vivo pharmacokinetics	dog	Gentamicin	Silicone Rubber	7-11
Antibiotic Delivery	in vitro in vivo pharmacokinetics	rat	Tetracycline	HEMA/MMA	12-15
Schistosomiasis	in vivo	mouse	Niridazole	Silicone Rubber	16

Sustained Release of Gentamicin From Prosthetic Heart Valves

Although prosthetic heart valves have been extensively and successfully employed clinically, problems still exist during the postoperative period concerning thromboembolism, infection and valvular incompetence. Infectious prosthetic valve endocarditis (PVE) remains as a serious and potentially devastating consequence of valve replacement surgery with mechanically fabricated prosthetic valves. Prophylactic antibiotic usage has reduced the clinically confirmed incidence of PVE but early mortality due to such infections remains high, 70% or greater as reported in several studies.

An earlier study has indicated the ability of parenteral prophylactic antibiotic therapy to reduce the incidence of prosthetic mitral valve failure in dogs associated with valve related infections. As an alternative to the above approach, we have investigated the therapeutic efficacy of a system providing for a sustained release of gentamicin, incorporating the antibiotic in silicone rubber which is in the sewing rim of a mitral valve prosthesis implanted in dogs.

<u>In vitro</u> studies with the uncoated gentamicin loaded valve rim inserts revealed an initial burst of drug released which peaked on the first day and slowly fell through day 10. At that time, a near constant pattern ensued during which 2.5-7.0 mg/day were released for a period of 2-3 wks followed by lower levels of drug release for up to 2 mos. Valve rim inserts with the additional drug-free silicone rubber layer (coated samples) exhibited a different release pattern with a reduced burst effect and greater persistence of long-term drug delivery. Peak release with the coated valve rim insert was delayed to the third or fourth day and after 4 days mean cumulative gentamicin release was only 59% of that of the uncoated inserts.

Serum measurements in dogs receiving the uncoated antibiotic loaded valve rims containing the ^{3}H-gentamicin exhibited a large early burst of gentamicin release, similar to that observed in the <u>in vitro</u> experiments. Peak serum gentamicin levels occurred during the cardiopulmonary bypass procedure. Serum drug concentrations rapidly fell over the first 3 to 4 days following valve implantation and remained at levels of approximately 0.2 µg/ml for the next 2 to 3 wks. Serum gentamicin concentrations were measurable for 1 to 2 mos before falling below detectable levels (< 0.1 µg/ml). The initial burst of gentamicin release was absent as measured in the serum of dogs implanted with prosthetic valves containing the coated rim inserts. The additional layer of drug-free silicone over the drug loaded valve insert effectively reduced peak serum gentamicin concentrations such that levels remained relatively constant over the observable drug delivery period.

Survival data on dogs implanted with mitral valves containing drug-free silicone rubber rim inserts (controls) and rim inserts loaded with 40% gentamicin sulfate, by weight, both uncoated and coated samples, are presented in Table II. A significant increase in survival is readily apparent for dogs receiving gentamicin loaded valves (uncoated and coated) compared to the animals receiving the control valves with the drug-free silicone rubber inserts. Long-term survival (dogs living 5 mos or greater) was 70% for the combined uncoated and coated gentamicin loaded valve implants, compared to only 31% in the control group. Three of the dogs in the gentamicin loaded valve rim group were alive 22 mos or more postimplantation.

Table II. Survival Data on Dogs Receiving Control and Gentamicin Loaded Prosthetic Mitral Valves

Group (Total # Animals)	Died Before 1 mo.	Alive After 1 mo.	Percent Survival After 1 mo.
Control Valves 9	6	3	33%
Uncoated Valves 24	4	20	83% ($p<0.05$)
Coated Valves 13	3	10	77% ($p<0.10$)

To investigate the in vivo efficacy of sustained-release preparations a comprehensive pharmacokinetic model is necessary. The two basic processes depicted within a given analysis are the primary diffusive transport of the active agent through the polymer delivery vehicle into the biological receiving medium and the subsequent transport, distribution and elimination of the drug. The latter process has been detailed in numerous pharmacokinetic models relating the observations on absorption, distribution, and elimination of orally or parenterally administered conventional pharmaceutical agents. The former process is unique to the topic of sustained drug release and is related to the physical and chemical characteristics of the drug and its polymer delivery vehicle and the biological response elicited by the implanted system in situ.

To couple the in vitro release rate behavior of the gentamicin releasing prosthetic valve system with observations on serum-gentamicin levels after implantation of the valve in dogs, an open one-compartment pharmacokinetic model was added to the release rate equations describing in vitro drug delivery. Gentamicin is not metabolized, and its distribution kinetics can be modeled adequately (within a 15% error) by a one-compartmental system. Elimination kinetics and the apparent volume of distribution in the individual dogs were derived from the gentamicin clearance studies.

Attempts were made to fit the in vivo serum concentration results to pharmacokinetic models based on revised diffusion-model equations for the uncoated and coated valve-rim inserts. To couple the in vitro diffusion equations to an in vivo simulation, a classical one-compartment model of drug distribution was employed. The mean apparent volume of distribution, V_D = 5.63 liters, and the first order eliminating constant, K_e = 0.695/hr, for gentamicin were determined in the intravenous injection-clearance studies.

Diffusion-model constants were recalculated using the results from in vitro flow-cell studies with suture punctured, Dacron mesh enclosed, valve-rim inserts. In general, all the model results fall short of predicting the actual serum-gentamicin concentrations determined experimentally over the observed valve implantation period.

From the results obtained in the attempts to model the in vitro release of gentamicin from the uncoated and coated valve rim inserts, it would appear that the actual drug-release process is more complex than predicted by relatively simplistic numerical analyses. For the uncoated rim insert system, computer fitting of the data was good over the early period of release but somewhat less adequate during the later time when a constant diffusion coefficient throughout the analysis was employed. The same trend was observed in the analysis of the coated rim insert system under identical conditions.

Release of gentamicin in vivo is an even more complex process. After implantation of the valve in the dog, the rate of release of the drug from the rim-insert surface would be dependent on such factors as: the relative amounts of surface exposed to the blood and the tissue, blood perfusion of the tissue, intracellular sequestration of the drug (both at the valve site and at peripheral organs), and the biological response of the blood and tissue layers adjacent to the valve surface. The fibrin pseudo-intima which forms early over the blood-contacting surface of the valve sewing ring might offer an additional diffusional resistance to the drug delivery device. At the tissue interface, fibrous capsule formation and the pooling of fluids within the surgically damaged tissue, or either one, may produce similar effects, modifying the release process.

In the later serum-gentamicin determinations, the elevation of measured serum-drug concentration over levels predicted by the pharmacokinetic model is most likely due to the inability of a one-compartment distribution model to properly describe the observed tissue sequestration of gentamicin and the later return of the drug from the tissues to the vascular compartment. This process would delay total drug excretion and cause an elevation of the later-time measurements of serum gentamicin concentration. Over the low-serum concentration range observed in the experiments after three to four days of valve implantation, the effects of tissue drug depots on serum gentamicin concentration would become significant.

Future pharmacokinetic models dealing with sustained drug release must take these effects into account through the use of time varying elimination parameters and multi-compartmental simulations. To this end, investigations are being conducted to better measure the tissue handling of gentamicin delivered from intramuscular sustained-release devices with different initial drug loadings and the effect of the cardiopulmonary-bypass procedure on drug-elimination characteristics.

In our bacteriology studies, it is apparent that parenteral antibiotic does provide some protection against the development of PVE when these results are compared to an earlier experiment which used prophylactic systemic antibiotic for a 10 day period following valve replacement. In addition, it appears that the gentamicin loaded valve inserts provide protection against the development of PVE. To better appreciate the pathogenesis of early-onset PVE and the therapeutic efficacy of the gentamicin releasing prosthetic valve, we are developing an experimental protocol which utilizes Staphylococcus aureus bacteremias and direct inoculation in the periprosthetic tissue.

These studies are important in that they demonstrate the absorption of gentamicin into the local tissue environment, periprosthetic cardiac tissue and sewing rim thrombi, following release from the device. In addition, they provide some indication as to the role that the local tissue environment plays in determining the serum and organ levels of gentamicin. The role that the healing phenomena in prosthetic valves may play in determining drug levels in adjacent tissues and the availability of the drug to the systemic circulation is important. We have carried out experiments using a similar gentamicin-silicone rubber release system in dog skeletal muscle and have demonstrated the temporal and spatial variations of gentamicin levels in muscle tissue surrounding rod implants. These studies provide corroborative support for the effect of the fibrous capsule on retarding the permeation and sequestering of gentamicin in the preprosthetic tissue.

Sustained Release of Tetracycline From HEMA/MMA Copolymers

Previous experience with a hydrogel matrix-drug delivery system led to the investigation of a membrane-controlled drug delivery device in which a hydrophilic polymer core is supersaturated with a given pharmaceutical. The core-drug mixture is coated with a drug-free polymer layer which, by its hydrophobic nature (relative to the core material), functions as the rate-limiting membrane in drug diffusion from the device. Membrane-enclosed drug delivery systems maintain a constant activity source of the permeating drug at the interface between the core and the membrane-coating material. The result is a constant drug release rate that is proportional to the concentration gradient established over the diffusion rate-limiting outer membrane.

A membrane-controlled drug delivery device was developed to release tetracycline at zero-order rates. The tetracycline delivery vehicle is a trilaminate disk consisting of core and coating membranes fabricated from a series of 2-hydroxyethyl-methacrylate and methylmethacrylate copolymers. Appropriate adjustment of the monomer composition ratio imparts a hydrophobic nature to the copolymer outer coating membrane (relative to the core material), which serves as the rate-limiting membrane in drug diffusion. The trilaminate disks demonstrated a zero-order tetracycline release over 4 months in vitro. The zero-order release rate was a function of the general device geometry, coating membrane thickness, disk surface area, level of core reservoir drug loading, and membrane coating copolymer composition. Permeability parameters of tetracycline diffusion through a series of 2-hydroxyethylmethacrylate-methylmethacrylate copolymer membranes were determined by a flux-lag time method. Equilibrium hydration values of these membranes also were determined. The ability of trilaminate 2-hydroxyethylmethacrylate-methylmethacrylate devices to release tetracycline at constant rates over a prolonged period offers unique therapeutic and investigational possibilities.

The trilaminate tetracycline-releasing disks were evaluated in vivo by implantation in the peritoneal cavities of female Sprague-Dawley rats. The implanted disks were coated with copolymer membranes of either 2:98 or 22:78 2-hydroxyethyl-methacrylate-methylmethacrylate composition. In vivo release rates of tetracycline were constant with time, as measured by the cumulative drug release from disks recovered over different implantation periods. Steady-state tetracycline release rates from the trilaminate devices--158 µg/day for the 22:78 2-hydroxy-ethylmethacrylate-methylmethacrylate copolymer-coated disks and 123 µg/day for the 2:98 2-hydroxyethylmethacrylate-methylmethacrylate-coated disks--were in excellent correlation with the steady-state total daily excretion of the drug (153 and 110 µg/day for the two systems, respectively) by the rats. Plasma tetracycline concentrations reached steady-state levels within 2-3 days of trilaminate disk implantation and remained constant over the 14-day implantation period. Steady-state tissue tetracycline concentrations averaged 10-12 µg/g in the bone; 3-4 µg/g in the liver, kidney and GI tract; 2-3 µg/g in the muscle; and <1 µg/g in the fat tissue. A flow-limited pharmacokinetic model was constructed to simulate in vivo tetracycline delivery from the trilaminate disks and the subsequent drug distribution. Model predictions of tissue and plasma tetracycline concentrations based on the experimentally determined zero-order release rates were in good agreement with experimental measurements.

The ability of trilaminate controlled-release disks fabricated from 2-hydroxyethylmethacrylate-methylmethacrylate copolymers to deliver tetracycline at a constant release rate when implanted in rats has been demonstrated. A full range of in vivo zero-order release kinetics is attainable through variations in the copolymer composition of the diffusion-controlling coating membrane of the trilaminate disk. The ability of such a device to deliver a pharmaceutical at a desired constant release rate over a long period is extremely valuable both as a therapeutic and an investigational tool. This system may be extended to the study of other drugs where long-term constant delivery rates are desired. A flow-limited pharmacokinetic model is presented that utilizes the ability of the trilaminate controlled-release device to deliver tetracycline at constant rates over a prolonged implantation period. Predictions based on this model are consistent with tissue tetracycline concentrations measured in experiments with these implant devices. The physiological pharmacokinetic model can be utilized to provide reliable predictions of tissue drug levels and is useful in studies of differential organ toxicities due to effects caused by the parent drug or its metabolites. The model also contributes information valuable for the design of proper drug dosing regimens and for the fabrication and placement of controlled drug delivery systems designed to deliver drugs to selected physiological target sites.

The coupling of the principles of controlled release from a zero-order rate device and long-term physiological pharmacokinetic modeling is a unique research concept and will be used in investigational systems where such drug delivery characteristics help elucidate physiological rate mechanisms.

A controlled tetracycline delivery device, consisting of a membrane enclosed trilaminate disc fabricated from a series of 2-hydroxyethylmethacrylate-methylmethacrylate copolymers, demonstrated the ability to deliver tetracycline at zero-order rates *in vitro* and *in vivo* in rats and was applied to study the pharmacokinetics of tetracycline in the pregnant rat. The trilaminate discs containing tetracycline were implanted in pregnant Sprague-Dawley rats on the eighth day of gestation. The animals were sacrificed on day 19, 20, and 21 of gestation in order to measure the distribution of the controlled release tetracycline in the maternal, fetal, and placental tissues. Constant plasma tetracycline levels were attained in both the maternal and fetal circulations after 4 to 5 days postimplantation of the trilaminate discs. Placental transfer of tetracycline appeared rapid and no partitioning of the drug was observed between the maternal and fetal plasma. Tetracycline levels did not differ significantly in the maternal and fetal soft tissues (liver, kidney, G.I. tract, muscle, placenta) as measured over the last 3 to 4 days of the animal's gestational period. Highest tetracycline concentrations were determined in the fetal bone samples. In addition, some accumulation of the drug occurred in the amniotic fluid. A flow-limited pharmacokinetic model was constructed to simulate the distribution of tetracycline, delivered at a constant rate from the trilaminate device, in the pregnant rats. Predictions of fetal growth and maternal and fetal tissue tetracycline concentrations were in good agreement with the experimental measurements. The ability of these copolymer systems to deliver tetracycline at zero-order rates over extended periods offers numerous potential therapeutic and investigational applications, especially where such drug delivery characteristics are beneficial to the elucidation of physiological rate mechanisms, as in the pregnant animal.

Sustained Release of Niridazole From Silicone Rubber

Niridazole was incorporated into rubber implants to investigate the potential of a sustained release of the drug in the therapy of murine Schistosomiasis mansoni. The infected animals (200 cercariae for 6 weeks) were randomly divided into three groups: one group received silicone rubber implants containing 50% by weight niridazole; a second group received blank silicone rubber implants with no drug; and the third group received no implants. Mortality 4 weeks later was in excess of 80% for animals with no implants or with the blank silicone rubber implants. In contrast, 10% mortality was observed in the mice receiving the niridazole-silicone rubber implants over a 10-week period. The worm burden in the niridazole-silicone rubber implant group was reduced at 10 weeks postimplantation by 77%.

The dramatic differences in mortality and morbidity between the niridazole-treated and control mice in our experiments are probably the result of reduction of worm burden to levels which can be tolerated by the animals. In addition, the low levels of niridazole diffusing from the implants may contribute to amelioration of disease by suppressing granuloma formation. As delayed hypersensitivity reactions in the host have been shown to play a key role in the immunopathogenesis of schistosomiasis, a low level of niridazole may help to alleviate the consequences of disease.

In this initial investigation, an intraperitoneal implantation site was chosen for the purposes of delivering the drug to the appropriate physiological target site, i.e., the mesenteric circulation. Other polymer-drug systems are under investigation for delivering niridazole and other antischistosomal drugs over a range of release rates sufficient to accomplish the combined tasks of reducing worm burdens and minimizing disease symptomatology.

Acknowledgments

The author gratefully acknowledges the efforts of his collaborators: Drs. L.S. Olanoff, R.D. Jones, A.A.F. Mahmoud, T. Koinis and F.S. Cross; and Messrs. J. Pelagalli and H. Niven. Partial support for the work summarized in this article was provided by the National Institutes of Health. The author extends his appreciation to the following companies which provided drugs used in these studies: Eli Lilly and Co., the Schering Corp., and Ciba-Geigy, Inc.

Literature Cited

1. Theeuwes, F. J. Pharm. Sci. 1975, 64, 1987.
2. Wise, D. L.; Rosenkrantz, H. in "Drug Delivery Systems", Gabelnick, H. L., Ed., DHEW Publication (NIH) 77-1238, 1976, p 252.

3. Brooke, D.; Washkuhn, R. J. *J. Pharm. Sci.* 1977, *66*, 159.
4. Borodkin, S.; Tucker, F. E. *ibid.* 1975, *64*, 1289.
5. Shell, J. W.; Baker, R. W. *Ann. Ophthamol.* 1974, *6*, 1037.
6. Zaffaroni, A. *Acta Endocrinol.*, *Suppl.* 1974, *185*, 423.
7. Olanoff, L. S.; Jones, R. D.; Anderson, J. M. *Trans. Amer. Soc. Artif. Internal Organs* 1979, *25*, 334.
8. Olanoff, L. S. Ph.D. Thesis, Case Western Reserve University; 1980.
9. Olanoff, L. S.; Anderson, J. M.; Jones, R. D. Pharmacokinetic Modeling of Gentamicin Release From a Prosthetic Heart Valve, in "Controlled Release of Bioactive Agents", Lewis, D. H., Ed., Plenum Press, 1981; in press.
10. Anderson, J. M., Niven, H., Pelagalli, J., Olanoff, L. S., and Jones, R. D. "The Role of the Fibrous Capsule in the Function of Implanted Drug-Polymer Sustained Release Systems" *J. Biomed. Mater. Res.*; in press.
11. Anderson, J. M.; Jones, R. D.; Pelagalli, J.; Niven, H.; Olanoff, L. S.; Cross, F. S. Prosthetic Valve Endocarditis and the Sustained Release of Gentamicin From Prosthetic Valves, *J. Thorac. Cardiovasc. Surg.*; submitted.
12. Olanoff, L.; Koinis, T.; Anderson, J. M. Controlled Release of Tetracycline I. *In vitro* Studies with a Trilaminate 2-Hydroxyethylmethacrylate-Methylmethacrylate System. *J. Pharm. Sci.* 1979, *68*, 1147-1150.
13. Olanoff, L.; Anderson, J. M. Controlled Release of Tetracycline II. Development of An *in vivo* Flow-Limited Pharmacokinetic Model, *J. Pharm. Sci.* 1979, *68*, 1151-1155.
14. Koinis, T. M.S. Thesis, Case Western Reserve University; 1980.
15. Olanoff, L. S.; Anderson, J. M. *J. Pharmacokin. Biopharm.*, 1980, *8*, 599.
16. Olanoff, L. S.; Mahmoud, A. A. F.; Anderson, J. M. *Am. J. Trop. Med. Hyg.* 1980, *29*, 71.

RECEIVED August 26, 1981.

Polymeric Delivery Systems for Macromolecules

Approaches for Studying In Vivo Release Kinetics and Designing Constant Rate Systems

ROBERT LANGER[1], DEAN S. T. HSIEH[1,2], and LARRY BROWN

Massachusetts Institute of Technology, Department of Nutrition and Food Science, Cambridge, MA 02139

> Methods for quantitating in vivo release kinetics and for designing constant-rate release systems for macromolecules are described. These systems are composed of ethylene-vinyl acetate copolymer and the incorporated drug. Approaches for directly quantitating in vivo release kinetics have been established using the polysaccharide, inulin. In vitro and in vivo release kinetics of inulin from the copolymer system were nearly identical. An approach for achieving zero-order release kinetics was tested by constructing matrices in the form of a hemisphere with all portions laminated with an impermeable coating except for a cavity in the center face. Zero-order release kinetics of bovine serum albumin were observed for 60 days.

Over the past 20 years research has been conducted in 2 primary areas concerning the use of polymers to improve the ways drugs are delivered to the body. One way has been to attach the drug to a polymer in order to impart a different affinity or specificity to the drug. The second approach has been to incorporate the drug inside a polymeric matrix to cause it to be slowly released into the body. In this paper, two aspects of the second approach are reviewed for polymeric systems for delivering large molecular weight (M.W. > 600) drugs.

Despite increasing advances in the use of controlled release polymeric delivery systems (1-6), it has, in general, proved difficult to develop such vehicles for the long-term administration of macromolecular drugs. This is largely due to the fact that most biocompatible polymers such as silicone rubber are impermeable to molecules over 600 molecular weight (7). Recent studies in our laboratory have demonstrated, however, that solvent casting of a variety of polymeric materials (ethylene-vinyl acetate

---

[1] Also affiliated with Children's Hospital Medical Center, Department of Surgery, Boston, MA 02115.

[2] Also affiliated with Harvard Medical School, Boston, MA 02115.

0097-6156/82/0186-0095$5.00/0
© 1982 American Chemical Society

copolymer, polyvinylalcohol, poly-2-hydroxyethyl-methacrylate in
the presence of powdered drug permits continuous release
of macromolecules for over 100 days (8). It appears that the
powders cause a series of interconnecting channels to be formed
in the polymeric matrix. These channels are large enough to
permit macromolecular diffusion but tortuous enough to cause a
slow and continuous release (9). We have also examined factors
affecting in vitro release kinetics (10) and have demonstrated
that the macromolecules released from the systems are biochem-
ically active in vitro and biologically active in vivo (11). The
polymers used have also been shown to be biocompatible in sensitive animal tissues (12). In this paper we review two areas of
our most recent research that provide further information on
these unique drug delivery systems. These are (1) the development
of techniques for comparing in vivo and in vitro release kinetics;
and (2) the development of approaches for achieving zero-order
release kinetics.

COMPARISON OF IN VITRO AND IN VIVO RELEASE KINETICS

In earlier studies in which macromolecules were released
from polymers in vivo (11), the substances tested were proteins
or DNA which were metabolized. Thus, a direct comparison of in
vivo and in vitro release was impossible. We therefore chose the
polysaccharide inulin which has a molecular weight of 5200 daltons.
Inulin was chosen as a marker because it is not metabolized, not
excreted by the glomerulus, and is neither reabsorbed or secreted
by the kidney tubules. It is not bound by plasma proteins and is
not toxic (13). Complete recovery of ^{3}H-inulin (14) or C^{14}-inulin
(15) in urine has been observed from intravenous injection into
rats, rabbits, dogs and humans. Thus, inulin provides an
excellent marker for comparing in vitro and in vivo release
kinetics because inulin recovered in urine can be directly compared to inulin collected in vitro.

Methods. Polymer slabs containing inulin were made as
follows: 1) ^{3}H-inulin was diluted with non-radioactive inulin to
a specific activity of 3.49 μCi/mg; 2) 586 mg of this inulin was
suspended in 15 ml of 5% ethylene-vinyl acetate copolymer in
methylene chloride; and 3) this suspension was cast in a square
7 cm x 7 cm x 0.5 cm glass mold, dried and cut into nine 1 cm x
1 cm squares as described previously (10).

In the in vitro kinetic experiments, 4 polymer squares were
placed in 5 ml of Phosphate Buffered Saline (PBS) at pH 7.4 and 37°C.
Polymer slabs were transferred daily to fresh PBS and the amount
of ^{3}H-inulin released was assayed as follows: a 250 μl sample of
PBS was added to 5 ml of Biofluour scintillation cocktail and
0.5 ml of distilled water. The emulsion was vortexed and counted
in a scintillation counter.

In the in vivo experiments, the remaining 5 squares were implanted subcutaneously in rats. The rats were housed in metabolic
cages and urine was collected daily. A 250 μl urine sample was

placed into 5 ml of Biofluour scintillation cocktail, vortexed and counted as above. Quenching of urine was corrected for using the channels ratio method (16).

Results. Excellent correlation between in vitro and in vivo release rates was observed (Figure 1). As in internal control, the inulin polymer squares were removed from the rats at the end of 450 hrs and rat urine analyzed 4.5 hrs later. As shown in Figure 1, the recovery rate of ^{3}H-inulin dropped 300 fold. This control confirmed that ^{3}H-inulin is rapidly cleared from the blood stream and excreted into the urine. Thus inulin is an excellent marker for directly assessing release rates from the polymer implant.

APPROACHES TO ACHIEVING ZERO-ORDER RELEASE KINETICS

Another limitation of the systems which were fabricated in the shape of a slab in earlier studies was that they released macromolecules at decreasing rates as a function of time. However, in many cases, it is desirable to have constant release kinetics so that the body will maintain a constant level of the released drug. The reason for the decreases in release rates observed in previous studies (10), may be explained by considering a typical implant, in the shape of a slab, as a model (17). Soon after implantation, the solid drug dissolves from the surface layer of the implant and then diffuses out of the implant. Since the drug doesn't have very far to travel to reach the surrounding media, release is rapid. As time elapses, the surface layer becomes depleted, so drug from deeper within the implant must dissolve and diffuse to the surface. The drug now has a longer distance to travel, so release rates decrease.

We thought it would be possible to compensate for this phenomenon if a polymer implant was designed in such a way that an increased area of drug would be available for release as the distance from the release surface increased. A variety of shapes were analyzed from a theoretical standpoint (18); the best results were obtained with a hemispheric device laminated with an impermeable coating, except for a small cavity in the center face (Figure 2). This can be envisioned as a small cantelope cut in half, where the orange pulp of the melon is the drug-polymer mixture. The melon half is then coated everywhere except where the seeds were, so all the drug must be released through the small exposed section. In this case although the drug must still travel increased distances at a later point in release time, this new shape compensates by providing more surface area of available drug as time elapses. To test this hypothesis, procedures were recently developed to construct hemispheric systems for the release of macromolecules (19) and these are discussed below.

Methods. The molds used for fabricating hemisphere shaped systems for macromolecules were composed of glass and had hemis-

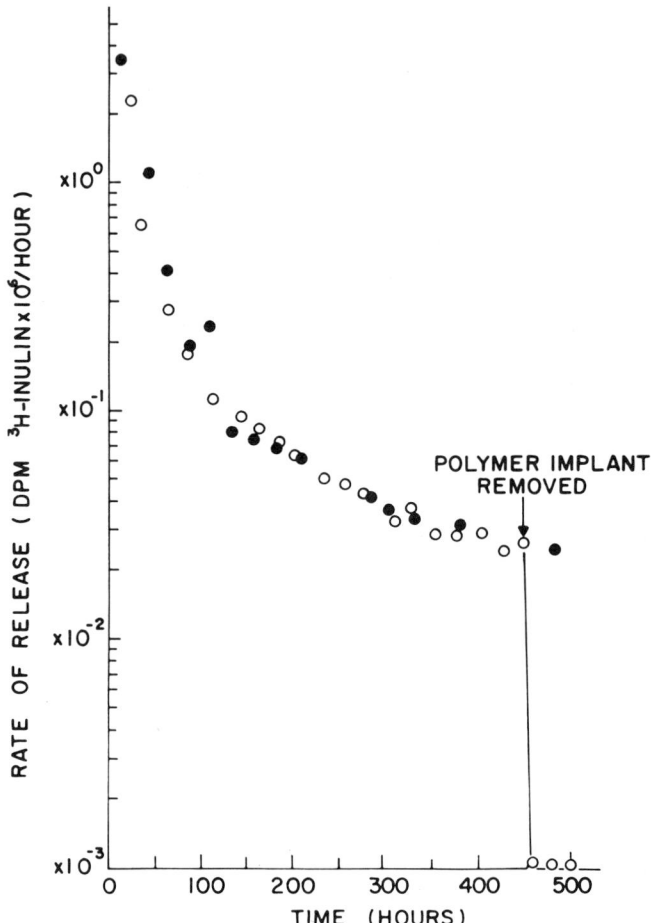

Figure 1. In vivo vs. in vitro comparison of rates of release of ^{3}H-inulin from ethylene–vinyl acetate copolymer at 44% loading. Key: implants into 5 rats; in vivo, $\bigcirc$, N = 5 polymers; and in vitro, $\bullet$, N = 4 polymers.

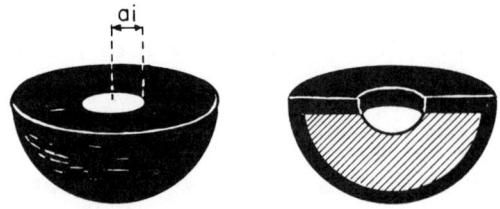

Figure 2. Schematic of an inwardly releasing hemisphere (a_i = inner radius). The hatched lines represent the dispersed zone of drug in polymer, and the black represents laminated regions through which release cannot occur. Key: top view, t = 0, (left); and side view cross section, t = 0 (right).

pheric bottoms (13 mm diameter x 11 mm height). These molds were made from glass test tubes by cutting off the bottoms. To prevent these molds from falling over, an embedding platform was made. This platform was constructed by pouring 15 ml of molten paraplast into a bacterial petri dish, and then placing empty glass molds into the paraplast so that they touched bottom. After allowing the paraplast to harden, the molds were removed and the indentations left behind in the paraplast could be used to subsequently support the molds.

To make the hemisphere polymer systems, earlier methods of preparing polymer-macromolecule slabs (10) were adapted. The procedure is illustrated in Figure 3 and described in detail as follows. First, blank glass molds were positioned in the indentations in the paraplast embedding platform. Then, 6 ml of 20% ethylene-vinyl acetate copolymer (EVA) solution made in methylene chloride and 514 mg of bovine serum albumin (BSA) powder (particle size range (10) from 150 to 180 μm) were mixed in a glass vial at room temperature and vortexed for 1 minute to yield a uniform suspension. Into each hemispheric glass mold, 0.8 ml of the protein-polymer dispersion were pipetted with a Pipetman. The petri dish containing these samples was then transferred onto a block of dry-ice for 10 minutes. The samples gelled within one minute. The petri dish containing these samples was then dried first at -20°C for two days and then at 20°C for another two days as reported previously (10).

The hemisphere pellets were twice coated with 20% EVA solution (containing no macromolecules) to form an impermeable barrier. The coating procedure is described as follows: (i) the tip of a cylindrical metal stick (1.8 diameter x 30 mm in length) was inserted into the center of the flat surface of each hemisphere pellet to a depth of approximately 3 mm. (ii) Using the uninserted portion of the stick as a handle, the hemisphere pellets were placed directly on the surface of a block of dry ice for 10 minutes. (The round bottoms of the pellets were touching the dry ice.) (iii) Again using the metal stick as a handle, the cooled hemisphere was then immersed into 20% EVA solution at 20°C for 10 seconds, removed, and then placed immediately on the same dry ice for 10 minutes. (iv) A second layer of coating was done in the same manner as described above. (v) The hemispheres were then put in the freezer (-20°C) for two days followed by further drying at 20°C for two days in a dessicator under a house-line vacuum (600 millitorr) to remove residual solvent (10). (vi) Finally, to create the exposed cavity in the face of the hemisphere pellet, the metal stick was removed by gently encircling the polymer surface immediately surrounding the stick with a scalpel-blade.

As controls, hemispheres of pure EVA were prepared uncoated and completely coated (including the cavity). Similarly, completely coated hemispheres containing BSA were prepared. All of these controls were tested for release kinetics.

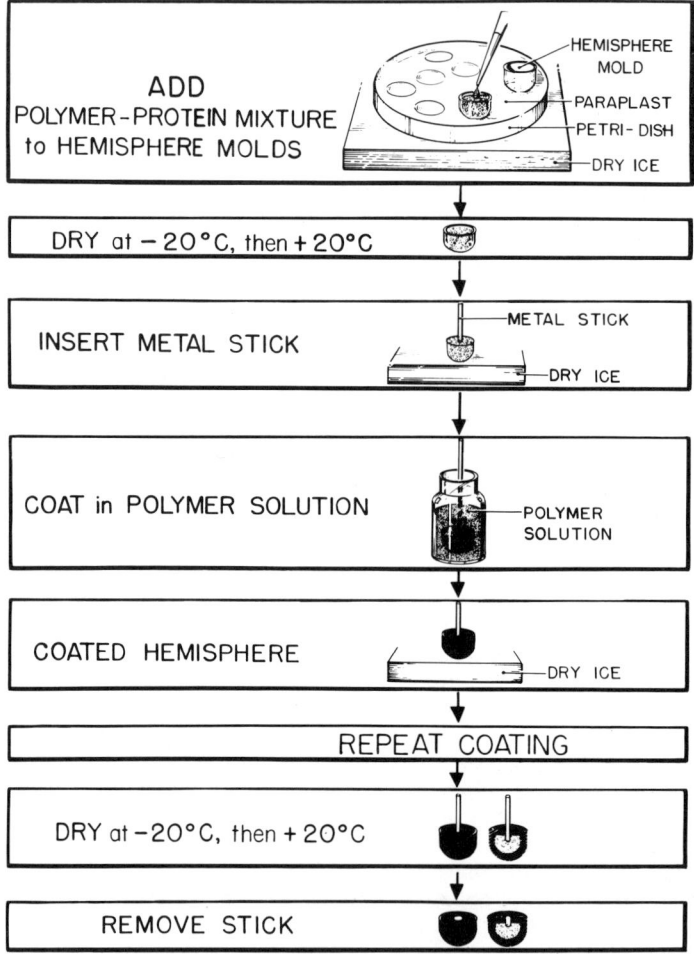

Figure 3. Flow sheet for preparing hemisphere-shaped devices for the release of macromolecules. In the last two blocks of the flow sheet, both the top view and a cross-sectional view of the same hemisphere device are shown. (See text for detailed procedures.)

The kinetics of BSA release from the hemisphere shaped systems were followed by methods described previously (10). BSA concentrations were measured spectrophotometrically by determining absorbance at 280 nm or 220 nm (19).

Results. Figure 4 shows the release kinetics of BSA from the hemisphere shaped systems. Each point represents 8 samples. A linear relationship between cumulative percentage release and the time of release was observed for 60 days. The release rate was approximately 0.5 mg of BSA per day. Standard error of the mean of cumulative release at each time point was within 12%. Control hemisphere shaped devices completely coated with 20% EVA solution, did not release any protein. The media collected in the release experiments of the control hemisphere shaped devices prepared from pure EVA solution (without BSA) showed no spectrophotometric absorbance.

DISCUSSION

The first set of experiments provides strong support for the assumption in earlier studies (8,9,10) that in vitro experiments would yield useful estimates for in vivo release kinetics. Further studies must still be conducted with other molecules and for longer times to fully validate the agreement between in vitro and in vivo release kinetics. Nevertheless, these studies demonstrate excellent agreement in the case studied here and establish a potentially important new technique for determining in vitro-in vivo release kinetics from polymeric delivery systems.

The second set of studies demonstrate that hemispheric matrices can act as constant release systems for high molecular weight compounds. The fabrication procedures for these systems are relatively simple and can be performed without expensive apparatus.

The release rate of macromolecules or proteins, such as BSA, from ethylene-vinyl acetate copolymer matrix slabs can be varied by as much as 2000-fold by manipulating three fabrication parameters, (i.e., drug loadings, aggregate sizes, and matrix coating) (10). The release rate of macromolecules from a hemisphere shaped device, can conceivably be changed by manipulating the same fabrication parameters and the size of the opening. In addition, we have previously demonstrated approaches for increasing the release rates of macromolecules from polymer slabs when desired by placing steel beads in the slabs and applying an external magnetic field (20). It is conceivable that a similar approach could be employed with the hemisphere devices to yield a constant release system supplemented by occasional bursts when needed (e.g., insulin delivery in diabetes). The design of such a system is currently an ongoing project in our laboratory.

Although the hemisphere design resulted in zero-order release kinetics for macromolecules such as BSA, the release mechanism for these systems has not yet been fully established. However,

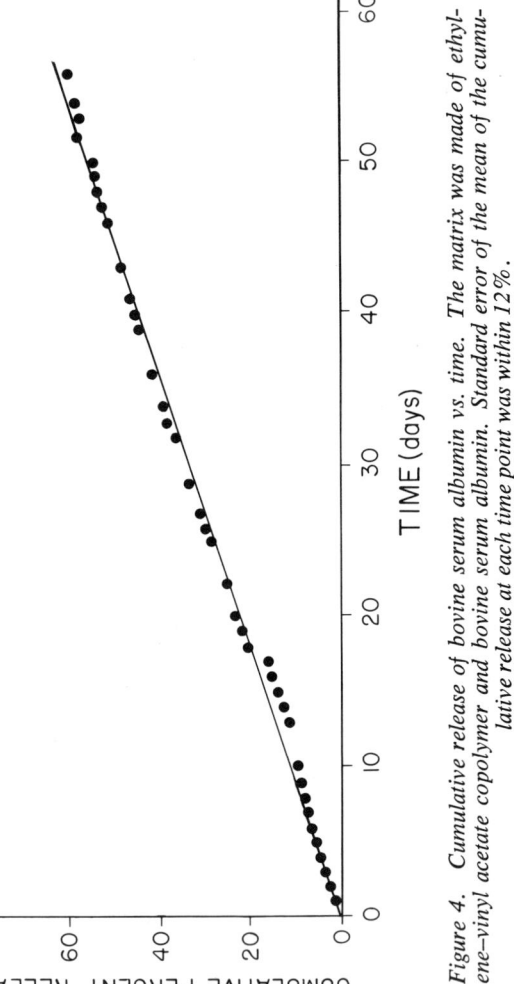

Figure 4. Cumulative release of bovine serum albumin vs. time. The matrix was made of ethylene–vinyl acetate copolymer and bovine serum albumin. Standard error of the mean of the cumulative release at each time point was within 12%.

evidence continues to indicate that the release of macromolecules from polymeric matrices occurs via diffusion through interconnecting pores (9). The present study is consistent with such a mechanism. However, a mathematical model that can be used to predict release kinetics from macromolecular release systems has not yet been developed.

Although only BSA was used as a model macromolecule in the second study, initial results have also indicated that zero-order release rates were obtained for over one month for lysozyme and β-lactoglobulin when they were incorporated into hemispheric matrices. The hemisphere systems have not yet been tested in vivo. However, future studies will explore the inulin model used in the first section of this paper.

The major problems encountered in the fabrication procedures in the present study are the techniques of both coating the matrix and opening the cavity uniformly. The development of a technique for opening the hole on the flat face of the hemisphere is very critical to achieve reproducibility. Besides the methods practiced in these studies, improved techniques to create holes on the flat face of the hemisphere shaped device should be explored. Techniques such as computer aided drilling may prove useful in future studies.

The delivery systems for macromolecules have been used in a number of biological and potential medical applications. These include:

- vehicles for bioassays of informational macromolecules (11);
- tools in biological studies of neovascularization (21);
- sources of chemical gradients in chemotactic studies (22);
- histochemical tools in brain research (23);
- insulin delivery systems (24);
- improved vehicles for immunization (25,26);
- interferon delivery systems (27); and
- delivery systems for anti-cancer agents (28).

In summary, the studies reported in this review provide 2 important demonstrations (1) that in vitro release kinetics of macromolecules such as inulin from ethylene-vinyl acetate copolymer matrices are identical to their in vivo release kinetics, and (2) that zero-order release for macromolecules can be achieved for over 60 days using a hemisphere design. Further experimentation in these areas should provide information that will be useful in the eventual design of controlled release systems for insulin and other important bioactive macromolecules.

Acknowledgments

This study was supported by N.I.H. Grant GM 26698 and grants from the American Diabetes Association and the Juvenile Diabetes Foundation. Dean S.T. Hsieh is a fellow of the Juvenile Diabetes

Foundation and is a recipient of a starter grant from the Pharmaceutical Manufacturer's Association.

References

1. Langer, R. Chem. Eng. Commun. 1980, 6: 1.
2. Langer, R.; Karel, M. Polymer News, in press.
3. Langer, R. Technology Review, 83: 26-34, 1981.
4. Robinson, J.R. Ed.; "Sustained and Controlled Release Drug Delivery Systems"; Marcel Dekker, New York, 1978.
5. Paul, D.R.; Harris, F.W. Eds.; "Controlled Release Polymeric Formulations"; American Chemical Society Symposium Series 33, Washington, 1976.
6. Tanquary, A.C.; Lacey, R.E. Eds.; "Controlled Release of Biologically Active Agents"; Vol. 47 of Advances in Experimental Medicine and Biology"; Plenum Press, New York, 1974.
7. Baker, R.W.; Lonsdale, H.K. "Controlled Release: Mechanisms and Rates"; in Controlled Release of Biologically Active Agents, Vol. 47 of Advances in Experimental Biology and Medicine, A.C. Tanquary; R.E. Lacey, Eds.; Plenum Press, New York, 1974, p. 15.
8. Langer, R.; Folkman, J. Nature, 1980, 263, 797.
9. Langer, R.; Rhine, W.; Hsieh, D.; Bawa, R. In "Controlled Release of Bioactive Materials", R. Baker, Ed.; Academic Press, New York, 1980, p. 83.
10. Rhine, W.; Hsieh, D.; Langer, R. J. Pharm. Sci., 1980, 69, 265.
11. Langer, R.; Folkman, J. in "Polymeric Delivery Systems, Midland Macromolecular Monograph 5"; R.J. Kostelnik, Ed.; Gordon and Breach, New York, 1978, p. 175.
12. Langer, R.; Brem, H.; Tapper, D. J. Biomed. Mat. Res., 1981, 15, 267.
13. Smith, H.W. Principles of Renal Physiology, Oxford University Press; New York, 1956, p. 73.
14. Gutman, Y.; Gottchack, C.W.; Lassiter, W.E. Science, 1965, 147, 153.
15. Cotbue, E. Fed. Proc., 1955, 14, 32.
16. Andreucci, V.E. Manual of Renal Micropuncture, Ideleson Publishers, Naples, Italy, p. 303-7.
17. Higuchi, T. J. Pharm. Sci., 1961, 50: 874.
18. Rhine, W.; Sukhatme, V.; Hsieh, D.; Langer, R. in "Controlled Release of Bioactive Materials"; R. Baker, Ed.; Academic Press New York, 1980, p. 177.
19. Hsieh, D.S.T.; Rhine, W.; Langer, R. J. Pharm. Sci., submitted.
20. Hsieh, D.; Langer, R.; Folkman, J.; Proc. Natl. Acad. Sci., 1981, 78, 1863.

21. Conn, H.; Berman, M.; Kenyon, K.; Langer, R.; Gage, J. Invest. Ophthal., 1980, 19, 362.
22. Langer, R,; Fefferman, M.; Gryska, P.; Bergman, K. Can. J. Microbiol., 1980, 26, 274.
23. Moskowitz, M.; Mayberg, M.; Langer, R. Brain Res., 1981, 212, 460.
24. Creque, H.; Langer, R.; Folkman, J. Diabetes, 1980, 29, 37.
25. Preis, I.; Langer, R. J. Immunol. Methods, 1979, 28, 193.
26. Langer, R,; (H. Van Vunakis; J.J. Langone, Eds.); "Immunological Techniques"; 1981, Methods in Enzymology, Academic Press: New York, 73, 57.
27. Langer, R,; Hsieh, D.S.T.; Brown, L.; Rhine, W. "Better Therapy With Existing Drugs: New Uses and Delivery Systems"; Biomedical Information Corporation, New York, in press.
28. Langer, R.; Brem, H.; Falterman, K.; Klein, M.; Folkman, J. Science, 1976, 193, 70.

RECEIVED July 7, 1981.

Design of Polymeric Iron Chelators for Treating Iron Overload in Cooley's Anemia

ANTHONY WINSTON, JAMES ROSTHAUSER, DAVID FAIR, JAMSHED BAPASOLA, and WEERASAK LERDTHUSNEE

West Virginia University, Department of Chemistry, Morgantown, WV 26506

Iron chelating polymers based on hydroxamic acid and catechol groups have been synthesized for use in iron chelation therapy for treating iron overload in Cooley's Anemia. A mouse screen for in vivo activity of iron chelators has shown that poly(N-methacryloyl-β-alanine hydroxamic acid) (P-11) is effective and about equal to the standard drug desferrioxamine. The 11-atom spacing between hydroxamic acid groups is about optimum for intramolecular iron chelation and good solubility of the iron complex. A similar polymer, poly(N-methacryloylhydroxamic acid) (P-3), which has only 3 atoms between hydroxamic acids, is also active, but also rather toxic, a result that can be attributed to the tendency of P-3 to form insoluble intermolecular crosslinked complexes. A polymer bearing catechol groups is weekly active in removing iron in the mouse screen. Improvement may be possible with better control of the spacing of the catechol groups.

Cooley's Anemia is a genetic disorder relatively rare in the United States, but found more often among peoples of the Mediterranean area and within a band extending through India and southeast Asia (1). In addition to severe anemia, the disease causes bone malformation and general deterioration and distruction of the vital organs leading eventually to death.

The disease is characterized by an inability of the body to synthesize adequate amounts of the β chain of hemoglobin. The usual and really the only treatment is to administer blood transfusions on a regular basis throughout the lifetime of the individual. However, such a continual whole blood transfusion raises the iron level to the point where iron deposits form in the liver, spleen, bone, heart, and other organs. Heart failure in the teens and early 20's is usually the final outcome (2).

In order to retard or prevent this accumulation of iron, a powerful iron chelator, desferrioxamine-B, is administered either by injection or by intravenous infusion. This chelator is capable of forming a very stable complex with iron and then carrying the iron out of the body with the urine or stool. Iron chelation therapy greatly improves the well-being of the patient and extends the lifetime of the patient appreciably (3).

Desferrioxamine-B (DFB) (I) is produced in a fermentation process by Ciba Pharmaceutical Co. and is supplied as the methane sulfonate salt under the name Desferal. Although there are many excellent iron chelators known, DFB is the only one that not only works in humans, but is available in sufficient amounts to supply the need for iron chelation therapy.

The iron chelating ability of DFB is due to the hydroxamic acid, a functional group that possesses a natural affinity for iron(III). Also inherent in this structure is a powerful chelate effect due to the 9 atom spacing between hydroxamic acid groups, which permits three neighboring hydroxamic acids to fit the octahedral coordination sphere of the iron(III) without severe steric strain.

The chelate effect of DFB causes the formation constant to be greater than that of the monomeric acetohydroxamic acid by two orders of magnitude, Table 1. Other naturally occurring iron chelators, called siderochromes, include various modifications of DFB, ferrichrome (II), rhodotorulic acid (III) and enterobactin (IV), all of which possess high formation constants for iron, due in part to powerful chelate effects, Table 1.

Table 1.
Formation Constants of Iron Chelators

Chelator	log K	Reference
Acetohydroxamic Acid	28.2	(4)
Desferrioxamine-B	30.6	(4)
Desferrioxamine-E	32.5	(4)
Desferrichrome	29.1	(4)
Rhodotorulic Acid (RA)	62.3[a]	(5)
Enterobactin	52	(6)

[a]$Fe_2(RA)_3$

Although DFB therapy is effective in removing large quantities of iron rapidly, there are some drawbacks in its use. One of these is the short plasma residence time, about 30 minutes, which causes a significant reduction in the efficiency of DFB to remove iron. To counteract this rapid plasma clearance, the chelator is often administered by means of a portable pump that is worn by the patient and which continuously administers a controlled amount of the drug (7). On the other hand, the pump

I

II

III

IV

is often psychologically unacceptable, and frequent injections throughout the day are painful and not very practical.

For these reasons a program designed to synthesize new iron chelators is being supported by the Cooley's Anemia Foundation of New York and the National Institute of Arthritis, Metabolism and Digestive Diseases of the National Institutes of Health. The object is to devise structures that not only remove iron from the body but also possess properties that provide a better mode of administration. A definite advantage would be achieved if the new drug could be given orally, the most convenient route. Also, if the plasma survival time could be extended, the efficiency of drug utilization would be increased, and thus the frequency of treatment and quantity of drug required would be reduced.

Advances have been made in the area of designing and synthesizing chemical structures for binding iron tightly and specifically, and possessing the appropriate solubility characteristics and chemical stability for drug use. On the other hand, there is no assurance that any such compound, no matter how carefully designed, will actually work in vivo. Although the compound must, of course, have a high affinity for iron, there is no simple relation between this property and the ability of the compound to remove excess iron from living systems (8). Adding to the complexity of the problem is the idea that there are several iron pools open to attack by an iron chelator, such as transferrin, ferritin and iron in transit. However, knowledge as to which of these pools is available to an iron chelator such as DFB is not understood. Also, there is evidence that the rat and mouse screens, used to provide a preliminary evaluation of the drug, do not always provide a measure of the behavior of the drug in humans. For example, rhodotorulic acid, which appeared promising on the basis of animal tests, nevertheless produced painful reactions when used in humans (9). Clearly there is much to be learned about iron transport, the manner in which iron chelators really work, and the effect of structure on adverse physiological reactions.

Design of Iron Chelators

The design of new iron chelators has been directed largely toward mimicking the naturally occurring siderochromes such as desferrioxamine and enterobactin. Three functional groups are under serious investigation, hydroxamic acids, catechols, and phenols. These functional groups have been incorporated into a variety of structures in order to produce compounds having exceptionally high stability constants for iron, close to, or higher than that of desferrioxamine. Unfortunately, a high iron binding constant does not assure high activity in vivo (8), and thus many compounds will have to be prepared and screened biologically in order to find a satisfactory balance of chemical and physiological properties.

Several powerful iron chelators, some recently synthesized specifically for the Cooley's Anemia program, are being tested to ascertain their potential as drugs for use in iron chelation therapy for treating iron overload in Cooley's Anemia. Some of the more promising ones include:

Ethylenediamine-N,N'-bis(2-hydroxyphenylacetic acid) (EDHPA) (V) (8)

Derivatives of N,N'-bis(2-hydroxybenzyl)ethylenediamine-N,-N'diacetic acid (HBED) (VI) (8),

1,3,5 -N,N',N''-tris(2,3-dihydroxybenzoyl)triaminomethylbenzene (VII) (10,11),

1,5,9-N,N'N''-tris(2,3-dihydroxybenzoyl)triazacyclotridecane (VIII) (10)

poly(N-methacryloyl-β-alaninehydroxamic acid) (P-11) (IX) (12).

All of these compounds have exceptionally high binding constants for iron.

Polymeric Iron Chelators

The structure of the hydroxamic acid polymer P-11 was designed to maximize the formation constant for iron chelation. The formation constants of a series of such polymers is directly related to the magnitude of the chelate effect, which is, in turn, a function of the side chain length, Table II. A maximum chelate effect appears at a spacing of 11 atoms. A shorter spacing, 9 atoms, (P-9) is insufficient to permit three neighboring hydroxamic acids to fit about a single iron without strain, hence the lower value of log K. With increasing side chain length, P-13 and P-15, the chelate effect decreases with the increasing degrees of freedom and a grandual decrease in log K results (12).

Table II.
Formation Constants of Iron Complexes of
Hydroxamic Acid Polymers (12)

Polymer[a]	n[b]	log K_{Fe}
P-9	1	28.6
P-11	2	29.7
P-13	3	29.3
P-15	-[c]	29.0

[a] Code number indicates number of atoms separating hydroxamic acid groups.
[b] Numbers refer to structure **IX**.
[c] Spacer is glycylglycine.

V

VI

VII

VIII

IX

Another hydroxamic acid polymer, poly(N-methylacryloyl-hydroxamic acid) (P-3) (**X**) in which the hydroxamic acid units are separated by only three atoms was similarly prepared. Because of the short side chains and closely spaced hydroxamic acid groups, P-3 tends to cross link and precipitate on complexation with iron. In cases such as this, the groups are too closely spaced to give soluble intramolecular complexes with iron.

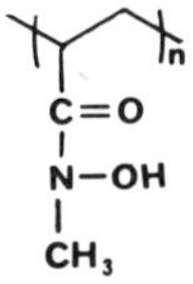

$$\begin{array}{c} \text{―(CH}_2\text{―CH)}_n\text{―} \\ | \\ C=O \\ | \\ N\text{―OH} \\ | \\ CH_3 \end{array}$$

X

Hydroxamic Acid Polymer P-11 has been prepared in two molecular weight varieties, $[\eta]_{DMF-H_2O}$ 2.24 (HP-11) and 0.87 dl/g (P-11). The formation constant reported in Table II is for the lower molecular weight sample. Although the formation constant of the higher molecular weight material has not been determined, it is expected that there would not be much difference between the two.

A polymer (P-DHB) (**XI**) based on catechol, the active functional group of enterobactin, was recently synthesized by the reaction of polyvinyl amine with the ethyl ester of 2,3-dihydroxybenzoic acid (DHB). Only about one third of the amine groups was found to be substituted with DHB units. The formation constant of the iron(III) complex (log K = 40) is the same as that reported for the simple dimethyl amide of DHB and so there does not appear to be any appreciable chelate effect.

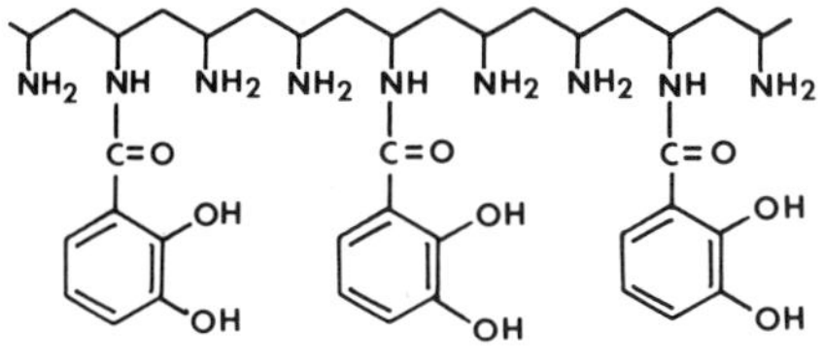

XI

Bioassay of Polymeric Iron Chelators

The iron chelator screen consists essentially of treating iron overloaded mice with the iron chelator by daily I.P. injections over a period of 7 days. Urine and stools are collected daily. At the end of the week, the mice are sacrificed and the livers and spleens are collected. The samples are then reduced to a suitable form and analyzed for iron by atomic absorption. An active chelator should produce increased iron levels in the urine and stool and decreased iron in the liver and spleen. Polymers P-11 (both molecular weight forms) P-3 and P-DHB were subjected to the mouse screen. The results are shown in Table III and compared with the drug DFB, which is used as a standard for comparison.

Overall, P-11 and HP-11 are judged to be about equal in potency to the standard DFB and except for lower fecal output occasionally observed at higher dose levels, both appear to be completely non-toxic. In the case of P-3, it is suggested that the extreme toxicity and high mortality is due to the tendency of P-3 to cross-link and precipitate on iron chelation. Such an event occurring in the blood stream would have obvious deleterious effects. Since the structural features of P-3 are similar to P-11 it is difficult to account for the toxicity on any basis other than the spacing of the groups. Regardless of the exact cause of the observed toxicity, it is clear that changes in spacing of the chelating groups profoundly affect both chemical and physiological activity.

P-DHB shows some weak activity in the mouse screen and gives some hope to the idea that a better placement of the DHB groups may result in a more positive biological outlook.

Potential Advantage of Polymeric Iron Chelators

Although the synthetic work on polymeric iron chelators has been directed primarily toward achieving strong chelate effects by appropriate spacing of functional groups, there is another means by which polymers might be of some distinct advantage over smaller molecules. And that concerns the residence time of the chelator in the plasma. As mentioned above, DFB is almost completely cleared from the plasma in 30 minutes, an effect which lowers the efficiency of DFB considerably. Although we do not know the direct cause of the rapid clearance of DFB, two possibilities have been suggested. Either the uncomplexed form is being degraded by metabolic processes, or the chelator is simply being lost by normal diffusion out through the membranes. Although degradation is a likely path, nevertheless, there is evidence to show that the iron complex of DFB is eliminated intact. If the loss of DFB is by simple diffusion, a high molecular weight analog of DFB, polymer P-11 for example, might diffuse out less rapidly, increasing the active lifetime in the

Table III.

Results of Mouse Bioassay of Polymeric Iron Chelators[a]

Treatment		Iron Content, % change			
Drug	Dose mg/kg	Spleen	Liver	Feces	Urine
		– – – – – First Series – – – – –			
None[b]	–	[+190]	[+206]	[−10]	[+33]
DFB	125	0	−15	+28	+181
	250	+3	−25	+76	+238
	500	+7	−35	+81	+450
P-11[c]	100	−17	−24	+44	+138
	500	−41	−49	−3	+331
		– – – – – Second Series – – – – –			
None[b]	–	[+594]	[+347]	[+9]	[+30]
DFB	250	−5	−22	+10	+498
HP-11[d]	500	−35	−12	−12	+631
P-3[e]	375	−46	−6	−80	+73
P-DHB[f]					

[a]Percent changes are based on the values of transfused control mice.
[b]Values in brackets compare tranfused controls with nontransfused controls.
[c]Decreased fecal output was observed at the 500 mg/kg dose level, otherwise no toxic signs were observed.
[d]No toxic signs observed.
[e]Severe toxic reactions, 6/10 mice died during the 7 day treatment.
[f]Weakly active as indicated by a 60% elevation of iron in the urine compared to control.

plasma and thus increasing the efficiency of iron removal by the hydroxamic acid complexes.

A dynamic factor involving changes in chain conformation on complexation might be involved. The uncomplexed polymer is highly solvated and would exist in an expanded state as a random coil. On complexation with iron, the polymer shrinks to occupy a smaller volume. When we consider that passage through membranes depends not only on the molecular weight, but also on the molecular volume, it is not unreasonable to suppose that a polymer of some select molecular weight might easily be passed by a membrane while in the complexed state, but be retarded or blocked by the membrane when in the form of a larger uncomplexed random coil. Such a polymer would circulate in the plasma until sufficiently complexed with iron to be small enough to diffuse out through the membranes.

Evidence that such polymers are capable of this dynamic behavior are the following observations. When an aqueous solution of a hydroxamic acid-acrylamide copolymer (containing 1-2 mole percent hydroxamic acid units) is treated with iron, the viscosity of the solution changes. At high polymer concentrations the viscosity increases dramatically, indicating an increase in molecular weight through the formation of intermolecular crosslinks involving iron-hydroxamic acid complexes. On the other hand, at low polymer concentrations, addition of iron causes a decrease in viscosity, clearly indicating a reduction in the molecular volume of the copolymer caused by intramolecular crosslink (13).

Iron complexation also affects the passage of the polymer through membranes. A low molecular weight variety of P-11 was dissolved in water and divided into 4 portions. Two of the portions were treated with standardized iron solution in an amount sufficient to form the 3:1 HA:Fe complex, and then placed in Spectrapor membrane tubes, one of 6000 and one of 12,000 mol. wt. cut off. The two other portions were diluted to the same extent with distilled water and similarly placed in two membrane tubes. After 24 hours the water surrounding the iron complexed samples had become red-brown, indicating passage of the polymer complex through the membrane. Samples of the water surrounding the membrane tubes containing uncomplexed polymers were treated with a few drops of the iron solution to form the colored complex of any polymer that had passed through the membrane. Some color was produced, but the color was not as intense as that from the iron complexed pair. Thus, the rate of diffusion of the polymer-iron complex through the membrane was greater than the rate of diffusion of the pure uncomplexed polymer. In principle, these results support the idea that a polymeric iron chelator could be designed to possess a long plasma residence time to collect iron and a relatively short residence time for the complex itself. On the other hand, since the actual processes and mechanisms of iron chelation therapy are largely unknown, whether or not such a drug really could be made to work in this fashion is difficult to say. But clearly further work is needed to clarify these points.

Acknowledgment. The direct support of this work by the Cooley's Anemia Foundation is greatly appreciated. The biological testing was performed by the Mason Research Institute, Worcester, Mass. and R. Grady, Rockefeller University, New York, under contracts with the National Institutes of Health, NIAMDD.

Literature Cited

1. Friedman, M. J.; Trager, W., Scientific American, 1981, 244 (3), 154.
2. Necheles, T. F.; Allen, D. M.; Finkel, H. E., "Clinical Disorders of Hemoglobin Structure and Synthesis", Appleton-Century-Crofts, N.Y., 1969.
3. Zaino, E. C.; Roberts, R. H., Editors, "Chelation Therapy in Chronic Iron Overload", Symposia Specialists, Miami, Florida, 1977.
4. Anderegg, G.; L'Eplattenier, F.; Schwarzenbach, G., Helv. Chim. Acta, 1963, 46, 1409.
5. Carrano, C. J.; Cooper, S. R.; Raymond, K. N., J. Amer. Chem. Soc., 1979, 101, 599.
6. Harris, W. R.; Carrano, C. J.; Cooper, S. R.; Sofen, S. R.; Ardeef, A. E.; McArdle, J. V.; Raymond, K. N., J. Amer. Chem. Soc., 1979, 101, 6097.
7. Propper, R. D.; Nathan, D. G., "The Use of Desferrioxamine and "The Pump"", in "Chelation Therapy in Chronic Iron Overload", Zaino, E. C.; Roberts, R. H.; Eds., Symposia Specialists, Miami, Florida, 1977, pp. 17-35.
8. Pitt, C. G.; Gupta, G.; Estes, W. E.; Rosenkrantz, H.; Metterville, J. J.; Crumbliss, A. L.; Palmer, R. A.; Nordquest, K. W.; Sprinkle Hardy, K. A.; Whitcomb, D. R.; Byers, B. R.; Arceneaux, J. E. L.; Gaines, C. G.; Sciortino, C. V., J. Pharm. and Exp. Therapeutics, 1979, 208, 12.
9. Grady, R. W.; Peterson, C. M.; Jones, R. L.; Graziano, J. H.; Bharguva, K. K.; Berdoukas, V. A.; Kokkini, G.; Loukopoulous, D.; Cerami, A., J. Pharm. and Exp. Therapeutics, 1979, 209, 342.
10. Harris, W. R.; Raymond, K. N., J. Amer. Chem. Soc., 1979, 101, 6534.
11. Venuti, M. C.; Rastetter, W. H.; Neilands, J. B., J. Med. Chem., 1979, 22, 123.
12. Winston, A.; Kirchner, D., Macromolecules, 1978, 11, 597.
13. Rosthauser, J. W.; Winston, A., Macromolecules, 1981, 14, 539.

RECEIVED July 7, 1981.

Structural and Functional Interrelationships of Anterior Pituitary Hormones

J. RAMACHANDRAN
University of California, Hormone Research Laboratory, San Francisco, CA 94143

The hormones of the anterior pituitary gland fall into three groups on the basis of structural similarities. Corticotropin (ACTH), the melanotropins (α-MSH and β-MSH), the lipotropins (β-LPH and γ-LPH) and endorphin (β-EP) comprise the first group of closely related flexible polypeptides. All the hormones of this group are derived from a single macromolecular precursor named proopiomelanocortin. Growth hormone (GH) and prolactin (PRL) are compact globular proteins with characteristic three-dimensional structure. These two hormones exhibit considerable sequence homology. The gonadotropins, lutropin (LH) and follitropin (FSH), as well as thyrotropin (TSH) are closely related glycoproteins forming the third group. These hormones are composed of two nonidentical subunits of comparable size. The α-subunit is common to the three glycoprotein hormones. Structure-function relationships of the pituitary hormones have been studied by the synthesis of analogs and by selective chemical modification.

The hormones of the anterior pituitary gland fall into three groups on the basis of structural similarities (1). The structural features of these hormones include all levels of organization of protein structure. Corticotropin (ACTH), the melanotropins (α- and β-MSH), lipotropin (LPH) and endorphin (EP) comprise the first group of flexible polypeptides in which information is organized in the linear sequence of amino acid residues. Growth hormone (GH) and prolactin (PRL) exhibit considerable structural homology and are compact globular proteins possessing characteristic three dimensional structure. The unique feature of the hormones of the third group is their quaternary structure. The three members of this group, namely, lutropin (LH), follitropin (FSH) and thyrotropin (TSH) are closely related glycoproteins composed of two non-identical subunits of comparable size. in the case of the hormones of groups 2 and 3, information is contained in the distinct conformation of the protein. Disruption of the three dimensional structure results in loss of biological activity.

Peptide Hormones

Structure-function relationships of the hormones in the first group have been investigated in detail by assessing the biological activities of numerous analogs of these hormones which have been synthesized. The rapid progress in the field of peptide synthesis in the late fifties and early sixties made it feasible to synthesize peptides ranging in size from 10 to 40 residues. The introduction of several new protecting groups and improvements in separation procedures have vastly increased the utility of solid-phase techniques for peptide synthesis. As a result, the total synthesis of all the hormones of the first group have been accomplished (2-6).

ACTH and MSH

ACTH is a single chain polypeptide composed of 39 amino acid residues (Figure 1). The amino acid sequence of α-MSH is identical to the sequence of the first 13 residues in ACTH. In α-MSH the amino and carboxyl terminals are blocked with an acetyl and amide group, respectively (Figure 1). The heptapeptide sequence Met-Glu-His-Phe-Arg-Trp-Gly present in ACTH and α-MSH is also contained in the β-MSH molecules isolated from the pituitary glands of several species (7). The primary physiological function of ACTH is the acute regulation of glucocorticoid production by the adrenal cortex and the long term maintenance of the adrenal cortex. The major recognized function of the melanotropins is their ability to stimulate the dispersion of melanin in amphibian melanophores. ACTH is also able to induce melanin dispersion. In addition, both ACTH and MSH stimulate lipolysis in rabbit adipose tissue. That the melanophore stimulating activity associated with purified ACTH is intrinsic to the hormone and not due to contamination with the melanotropins, was conclusively demonstrated by the synthesis of a nonadecapeptide corresponding to the first 19 residues of ACTH (8). This synthetic peptide stimulated steroidogenesis in the adrenal gland and also elicited melanin dispersion.

Studies with synthetic analogs of ACTH and MSH provided an understanding of the manner in which information is organized along the linear sequence of amino acid residues (9, 10, 11). Inspite of the high degree of structural similarity between ACTH and α-MSH, the biological peoperties of the two hormones are highly specific. α-MSH has less than 0.04% of the potency of ACTH in stimulating glucocorticoid production in the adrenal gland. Conversely, ACTH is only 1% as potent as α-MSH in

α-MSH: Ac-Ser-Tyr-Ser-Met-Glu-His-Phe-Arg-Trp-Gly-Lys-Pro-Val-NH$_2$
　　　　 1　　　　　　　 5　　　　　　　　　　 10　　　　　 13

β-MSH:　 Asp-Glu-Gly-Pro-Tyr-Lys-Met-Glu-His-Phe-Arg-Trp-Gly-
　　　　 1　　　　　　　 5　　　　　　　　　 10

　　　　 Ser-Pro-Pro-Lys-Asp
　　　　　 15　　　　 18

ACTH:　　 Ser-Tyr-Ser-Met-Glu-His-Phe-Arg-Trp-Gly-Lys-Pro-Val-
　　　　 1　　　　　　 5　　　　　　　　　 10

　　　　 Gly-Lys-Lys-Arg-Arg-Pro-Val-Lys-Val-Tyr-Pro-Asn-Gly-
　　　　　 15　　　　　　　　 20　　　　　　　　　 25

　　　　 Ala-Glu-Asp-Glu-Ser-Ala-Glu-Ala-Phe-Pro-Leu-Glu-Phe
　　　　　　　　 30　　　　　　　 35　　　　　　　　 39

Figure 1. Structure of corticotropin and the melanotropins.

eliciting melanin dispersion in the melanophores of nonmammalian animals. The minimal structural features required for eliciting either response appear to be the same. The pentapeptide sequence His-Phe-Arg-Trp-Gly has been shown to stimulate both adrenocortical cells (12) and melanophores (13) to the same maximal degree as the intact hormones, although nearly million-fold higher concentration of the pentapeptide was required. It appears that the same pentapeptide segment of both ACTH and α-MSH is involved in functional interaction of the hormones with their respective target cells. Specificity of interaction is conferred by the amino acid residues attached to either side of the pentapeptide segment. The decapeptide ACTH[1-10] which includes the His-Phe-Arg-Trp-Gly segment is fully active as a steroidogenic agent at high concentrations whereas ACTH[11-19NH$_2$] is inactive even at very high concentrations. The nonadecapeptide ACTH[1-19NH$_2$] is equipotent to the intact hormone. These results imply that linking the functional ACTH[1-10] peptide to the ACTH[11-19NH$_2$] sequence serves to lower the concentration of the peptide required for full activity by several orders of magnitude. The linear sequence corresponding to residues 11-19 of ACTH serves as an attachment site. The interaction of this region of the hormone with the molecular surface of the receptor increases the local concentration of the functional segment of the peptide in the microenvironment of the receptor. α-MSH which lacks the strongly basic amino acid sequence Lys-Lys-Arg-Arg (residues 15-18 of ACTH) is unable to attach efficiently to the adrenocortical receptor at physiological concentrations and therefore does not stimulate the adrenocortical cell even though it contains the information necessary for activating the receptor. The structure of α-MSH is designed for productive interaction with the melanophore. Desacetyl α-MSH is only 25% as potent as α-MSH. The lack of the acetyl group at the amino terminal and the presence of an extended peptide chain beyond residue 13, hinder the interaction of ACTH with melanophores and, therefore, at physiological concentrations, only α-MSH is able to activate melanin dispersion in the melanophores.

Further studies have shown that although the same amino acid sequence is involved in the activation of receptors on adrenocortical cells and melanophores, there are significant differences in the structural features of these receptors even in the region involved in interaction with the active pentapeptide sequence (14). Selective chemical modification of the single tryptophan residue in ACTH by reaction with o-nitrophenylsulfenyl chloride caused profound alterations in the biological properties of the hormone (15-19). The analog, o-nitrophenylsulfenyl-Trp^9-ACTH (NPS-ACTH) was found to be inactive as a lipolytic agent on rat adipocytes (15) and only 1/100 as potent as ACTH in stimulating steroidogenesis in rat adrenocortical cells (18). NPS-ACTH was found to be a potent anagonist of ACTH-induced lipolysis (15) as well as adenylate cyclase stimulation in both rat adipocytes (17) and rat adrenocortical cells (18, 19). On the other hand, NPS-ACTH was 3.4 times as potent as ACTH in stimulating amphibian melanophores (16) and 4.4 times as potent as ACTH in stimulating lipolysis in rabbit adipocytes (14). Similar selective modifications of the Trp residue in α-MSH converted the hormone into a more potent melanin dispersing agent (16) and lipolytic agent in rabbit adipocytes (14). These results suggest that the complementary region of the specific ACTH receptor involved in interacting with the His-Phe-Arg-Trp-Gly segment must be structurally different from the complementary region of the MSH receptor involved in the interaction with the same peptide segment. Furthermore, the receptors on rat adipocytes and rat adrenocortical cells appear to be very similar and highly specific for ACTH. The receptors on amphibian melanophores and rabbit adipocytes resemble each other closely and are specific for MSH. Comparison of the steroidogenic, lipolytic and melanophore stimulating activities of a number of analogs of ACTH and MSH confirmed the above findings (11).

LPH and EP

Lipotropin was discovered fortuitously during the investigation of a simplified isolation procedure for ACTH (20). Two peptides exhibiting lipolytic activity in rabbit adipose tissue were isolated from ovine pituitaries and were designated β- and γ-LPH. Ovine β-LPH (21) is a linear polypeptide composed of 91 amino acid residues (Figure 2). The 58-residue γ-LPH is identical to the sequence region 1-58 of β-LPH. Residues 41-58 in LPH correspond to the sequence of β-MSH. The biological activities of β- and γ-LPH are similar to those of β-MSH and can be accounted for by the presence of the β-MSH sequence within the LPH molecule. The physiological role of LPH has remained obscure. Recent evidence strongly suggests that the biological function of LPH may be that of a prohormone for the opioid peptide β-endorphin.

The discovery of endogenous opiate peptides has been one of the most exciting developments in the peptide hormone field. In December 1975, Hughes, Kosterlitz and collaborators (22) reported the isolation, structure and synthesis of two similar pentapeptides from porcine brain with potent opiate agonist activity. They recognized that the entire peptide sequence of one of these peptides, namely, methionine enkephalin, is present in residues 61-65 of the lipotropin molecule (Figure 2). In their search for LPH from camel pituitaries, Li and Chung (23) isolated a 31-amino acid residue peptide, which corresponded to residues 61-91 of LPH. This peptide was found to be more potent than the enkephalins in its analgesic properties as well as in its affinity for brain opiate receptor, and was named endorphin (EP). β-EP has been isolated from porcine (24), ovine (25), bovine (26), rat (27) and human (28) pituitaries. The structure of β-EP appears to be highly conserved. Both camel (29) and human (30) β-EP

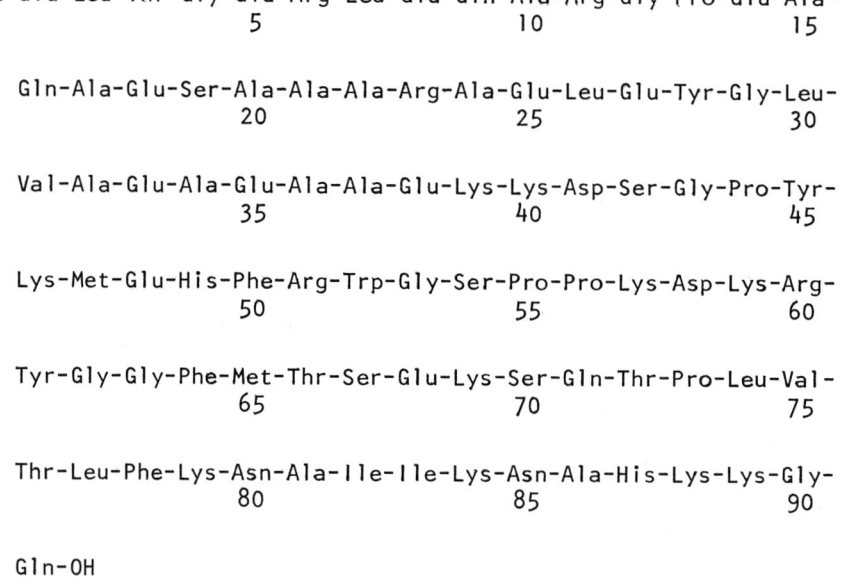

Figure 2. Structure of LPH. Endorphin corresponds to residues 61-91.

have been synthesized, in addition to a large number of analogs (31-35). Analysis of the structure-function relationships using the guinea pig ileum assay showed that the potency increases as the peptide chain length is increased from that of enkephalin to endorphin. The complete amino acid sequence of β-EP was found to be necessary for eliciting full analgesic activity in vivo. Blocking of the amino terminal of β-EP by acetylation results in loss of analgesic activity. This explains why β-LPH which contains the entire β-EP sequence has no analgesic activity. In addition to its analgesic activity, EP exhibits a remarkable array of biological properties modulating behaviour. Endorphin appears to be the most potent peptide in all biological tests. Structure-function studies using diverse biological assays may lead to the separation of these biological activities. Recent clinical studies indicate that synthetic EP exhibits beneficial effects in patients with severe pain, narcotic abstinence, schizophrenic behaviour and depression. Endorphin appears to interact with a variety of receptors in the brain. As a result of the enormous interest in brain peptides, it has become apparent that the other pituitary hormones such as ACTH and α-MSH are also present in discrete regions of the brain and that the hormones of this group may be involved in complex interactions in different areas of the brain, thus effecting modulation of behaviour.

Concurrent with the discovery of EP, it was recognized that all the hormones of the first group are formed biosynthetically by posttranslational processing of a high molecular weight precursor (36, 37, 38). The precursor, proopiomelanocortin, has a molecular weight of 31,000. The complete amino acid sequence of proopiomelanocortin has been deduced from the nucleotide sequence of the complimentary DNA (39). Lipotropin forms the carboxyl terminal and is preceded by the ACTH sequence. ACTH and LPH are separated by the paired basic residue sequence Lys-Arg. The amino terminal region also contains a sequence of residues common to both ACTH and LPH, namely His-Phe-Arg-Trp. In common with other protein precursors, the biologically active residues in proopiomelanocortin are flanked by paired basic residues at which proteolytic processing takes place.

Protein Hormones

The biological activities of the hormones of groups II and III are dependent on the unique conformations of these molecules. In this case, information is not confined to any linear segment of the polypeptide chain but is derived from the noncovalent interactions of distant segments of the polypeptide chain. Owing to the large size of these hormones structure-function studies have depended largely on selective chemical modification of one or more amino acid side chains. Studies with human growth hormone (hGH) provide a good illustration of this approach.

Growth Hormone (GH)

Human GH is a single chain polypeptide composed of 191 amino acid residues cross-linked by two disulfide bridges between cysteine residues at 58 and 165, and 182, and 189, respectively (Figure 3). Reduction of both disulfide bridges and alkylation with iodoacetamide yields the tetra-S-carbamidomethylated hGH which retains full biological activity in animals and humans, and its secondary and tertiary structure are indistinguishable from the structure of the intact hormone (40).

It was discovered recently that limited proteolytic digestion of hGH with human plasmin did not alter the biological properties of the hormone (41, 42). Li and Gráf (43) found that plasmin caused the excision of a hexapeptide corresponding to residues 135-140. The fragments 1-134 and 141-191 are held together by the disulfide bridge between cysteine 53 and 165 in plasmin-digested hGH. Reduction and alkylation of plasmin-digested hGH severs the covalent link between the two peptide fragments and they can then be separated from each other by gel filtration (43). Whereas plasmin-digested hGH is as potent as the intact hormone, the separated fragments have very little biological activity. However, recombination of the two fragments results in the formation of a noncovalent complex which is indistinguishable from the unmodified hormone in terms of biological activity as well as conformation as shown by circular dichroic spectra (44).

The dissociation of tetra-S-carbamidomethylated plasmin-digested hGH has been investigated and found to be a slow, exergonic equilibrium reaction (45). The major products of the dissociation have been shown to be the monomeric form of hGH(141-191) and a trimeric form of hGH(1-134). The carboxyl terminal fragment appears to be a random coil, whereas the amino terminal fragment, hGH(1-134) retains a considerable degree of secondary and tertiary structure. These results suggest that the noncovalent interaction between the fragments is essential for maintenance of the biologically active conformation of the hormone. Since hGH(1-134) by itself has about 14% activity and the activity of hGH(141-191) is negligible, it is likely that hGH(1-134) contains much of the information for favorable interaction with the GH receptor and hGH(141-191) probably stablizes the biologically active conformation of hGH(1-134).

An important consequence of these findings is that it is now possible to investigate structure-function relationships of hGH through semi-synthesis. Li et al. (46) showed that combination of the synthetic peptide corresponding to hGH(141-191) with the natural hGH(1-134) resulted in the full restoration of biological activity and immunoreactivity. Synthetic analogs containing norluecine in place of methionine and alanine in place of cysteine in hGH(141-191) as well as in hGH(145-191) all complemented with natural hGH(1-134) to generate full growth promoting activity

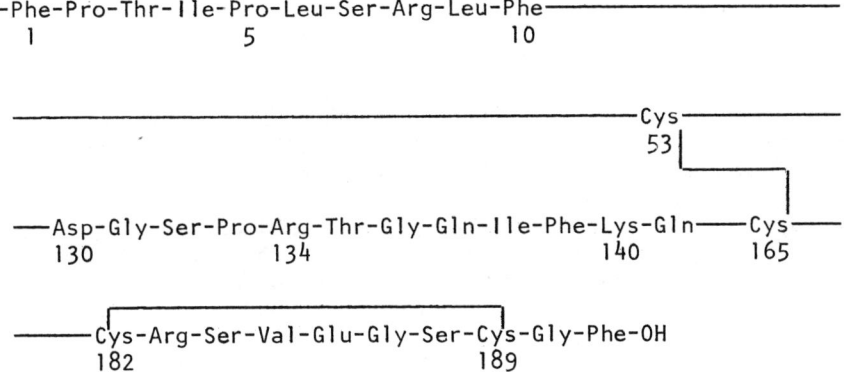

Figure 3. Structure of human growth hormone.

(47, 48). The importance of various amino acid residues in the fragment hGH(141-191) can be examined systematically by the semisynthetic approach.

Lactogenic Hormones

The primary protein hormone of the pituitary gland involved in regulation of lactation in mammals is prolactin (PRL). In addition to this, in humans there are two other hormones with lactogenic activity. Both hGH and human chorionic somatomammotropin (hCS) have identical lactogenic activities in the standard pigeon crop sac assay but their potencies are considerably lower than that of PRL. In the rabbit and mouse mammary gland assays, however, the lactogenic potencies of the three hormones are comparable. Comparison of the amino acid sequences of ovine PRL, hGH and hCS shows that there is considerable structural homology (49). Among all the known growth hormones, hGH alone has both growth-promoting and lactogenic activities.

PRL has also been selectively cleaved by limited proteolysis with fibrinolysin (50). This results in the cleavage of the peptide chain between residues 53 and 54. The two fragments PRL(1-53) and PRL(54-199) are devoid of both lactogenic activity and immunoreactivity. Recombination of the fragments resulted in full restoration of immunoreactivity but not biological activity. The physicochemical properties of the recombinant molecule were very similar but not identical to those of the native hormone. These results emphasize the enormous importance of the integrity of the conformation of these hormones for their biological functions.

Human CS, the placental hormone, is very similar to hGH in its primary structure (96% homology) yet has very low growth-promoting activity. Plasmin treatment of hCS results in the removal of a heptapeptide from the hormone (51). Reduction and alkylation of plasmin treated hCS gave two fragments corresponding to residues 1-133 and 141-191 of hCS. The fragments were devoid of any biological or immunochemical activity. Complementation of the carboxyl terminal fragment, hCS(141-191) with hGH(1-134) yielded a product with 50% growth promoting activity and nearly full immunoreactivity (52). The other hybrid, made up of hCS-(1-133) and hGH(141-191) had very low growth promoting activity (<10%). The immunological and receptor-binding properties of the two hybrids have been analyzed in detail (53). These studies have shown that the determinants for binding hGH to hepatic GH receptors as well as lactogenic receptors are present in the hGH(1-134) fragment of the hormone.

Glycoprotein Hormones

The gonadotropins LH and FSH, as well as TSH are all composed of two non-identical, non-covalently complexed, glycopeptide subunits. One of the subunits, designated α, is common to all three hormones. The β subunit of a glycoprotein hormone, therefore, determines the hormonal specificity. Individual subunits normally have little or no intrinsic biological activity (54). There is extensive structural homology (40-45%) between the β subunits. The glycoprotein hormones are characterized by a carbohydrate content of 15-25% and a high cystine content. All α subunits possess five disulfide bridges and β subunits possess six. LH has a high proline content, FSH is rich in acidic amino acids and TSH has a high tyrosine content. The placental hormones, human chorionic gonadotropin (hCH) and equine chorionic gonadotropin (eCG or PMSG) are structurally very similar to the pituitary glycoprotein hormones.

Selective chemical modification and recombination of the α and β subunits of the glycoprotein hormones to give intra and interspecies hybrids are the two major experimental approaches employed in elucidating the structural and functional interrelationships of this group of hormones. The alteration of an average of one histidine residue out of six in ovine LH by Rose Bengal-sensitized photooxidation resulted in 90% loss of biological activity (55). Based on a variety of chemical modifications of LH including nitration of tyrosine residues and acylation of the ε-amino groups of lysine residues, Ward (56) has proposed a model for the location of various functional groups involved in intersubunit contact and in the interaction with the receptor. In the α subunit the tyrosine in position 92 or 93, or both are considered essential for biological activity. Ward concludes that the α subunit accounts for the majority of the receptor-site binding, but the β subunit specifies the α subunit conformation and possibly one-third of the receptor-site binding specificity (56).

The role of the carbohydrate moiety in the actions of the glycoprotein hormones has been investigated by specific removal of carbohydrate by enzymatic (57) and chemical (58) methods. Chemical deglycosylation of ovine LH by treatment with anhydrous HF for 75 min at 0° resulted in the removal of nearly two-thirds of the carbohydrate moiety (58). Deglycosylated LH (DG-LH) was completely different from the native hormone in its physicochemical and biological properties. DG-LH could bind to LH receptors but did not stimulate cAMP production. The steroidogenic response of DG-LH was significantly attenuated compared to LH. Furthermore, DG-LH was found to be an effective antagonist of LH both in vitro and in vivo. These results suggest that the carbohydrate groups of the glycoprotein hormones are important for biological activity and that alteration of the carbohydrate may lead to antagonists which may prove useful agents for contraception.

Acknowledgment

The author wishes to thank Professor C. H. Li for his interest.

Literature Cited

1. Ramachandran, J. in "Metabolic Inhibitors"; Hochster, R. M.; Kates, M; Quastel, J. H. (eds.); Academic Press: New York, 1972; Vol. III, p. 361.
2. Blake, J; Li, C. H. Int. J. Peptide Protein Res. 1972, 3, 185.
3. Wang, K-T.; Blake, J.; Li, C. H. Int. J. Peptide Protein Res. 1973, 5, 33.
4. Yamashiro, D; Li, C. H. J. Am. Chem. Soc. 1973, 95, 1310.
5. Li, C. H.; Lemaire, S.; Yamashiro, D.; Doneen, B. A. Biochem. Biophys. Res. Commun. 1976, 71, 19.
6. Yamashiro, D; Li, C. H. J. Am. Chem. Soc. 1978, 100, 5174.
7. Ramachandran, J.; Li, C. H. Adv. Enzymol. 1967, 29, 391.
8. Li, C. H.; Meienhofer, J.; Schnabel, E.; Chung, D.; Lo, T. B.; Ramachandran, J. J. Am. Chem. Soc. 1960, 82, 5760.
9. Ramachandran, J. in "Hormonal Proteins and Peptides 2" Li, C. H. (ed.); Academic Press: New York, 1973, p. 1.
10. Schwyzer, R. Ann. N. Y. Acad. Sci. 1977, 297, 3.
11. Ramachandran, J; Farmer, S. W.; Liles, S.; Li, C. H. Biochim. Biophys. Acta. 1976, 428, 347.
12. Schwyzer, R.; Schiller, P.; Seelig, S.; Sayers, G. FEBS Lett. 1971, 19, 229.
13. Otsuka, H.; Inouye, K. Bull. Chem. Soc. Japan. 1964, 37, 289 and 1465.
14. Ramachandran, J.; Lee, V. Biochim. Biophys. Acta. 1976, 428, 339.
15. Ramachandran, J.; Lee, V. Biochem. Biophys. Res. Commun. 1970, 38, 507.
16. Ramachandran, J. Biochem. Biophys. Res. Commun. 1970, 41, 353.
17. Ramachandran, J.; Lee, V. Biochem. Biophys. Res. Commun. 1970, 41, 358.
18. Moyle, W. R.; Kong, Y. C.; Ramachandran, J. J. Biol. Chem. 1973, 248, 2409.
19. Ramachandran, J.; Kong, Y. C.; Liles, S. Acta Endocrinol. 1976, 82, 587.
20. Birk, Y.; Li, C. H. J. Biol. Chem. 1964, 239, 1048.
21. Li, C. H.; Barnafi, L.; Chrétien, M.; Chung, D. Nature. 1965, 208, 1093.
22. Hughes, J.; Smith, T. W.; Kosterlitz, H. W.; Fothergill, L. A.; Morgan, B. A.; Morris, H. R. Nature. 1975, 258, 577.
23. Li, C. H.; Chung, D. Proc. Natl. Acad. Sci. USA. 1976, 73, 1145.
24. Bradbury, A. F.; Smyth, D. G.; Snell, C. R. Biochem. Biophys. Res. Commun. 1976, 69, 950.
25. Chrétien, M.; Benjannet, S.; Dragon, N.; Seidah, N. G.; Lis, M. Biochem. Biophys. Res. Commun. 1976, 72, 472.
26. Li, C.H.; Tan, L.; Chung, D. Biochem. Biophys. Res. Commun. 1977, 77, 1088.

27. Rubinstein, M.; Stein, S.; Udenfriend, S. Proc. Natl. Acad. Sci. USA. 1977, 74, 4969.
28. Li, C. H.; Chung, D.; Doneen, B. Biochem. Biophys. Res. Commun. 1976, 72, 1542.
29. Li, C. H.; Lemaire, S.; Yamashiro, D.; Doneen, B. A. Biochem. Biophys. Res. Commun. 1976, 71, 19.
30. Li, C. H.; Yamashiro, D.; Tseng, L-F.; Loh, H. H. J. Med. 1977, 20, 325.
31. Yamashiro, D.; Tseng, L-F.; Doneen, B. A.; Loh, H. H.; Li, C. H. Int. J. Peptide Protein Res. 1977, 10, 159.
32. Yamashiro, D.; Li, C. H.; Tseng, L-F.; Loh, H. H. Int. J. Peptide Protein Res. 1978, 11, 251.
33. Yeung, H. W.; Yamashiro, D.; Chang, W-C.; Li, C. H. Int. J. Peptide Protein Res. 1978, 12, 42.
34. Li, C. H.; Chang, W-C.; Yamashiro, D.; Tseng, L-F. Biochem. Biophys. Res. Commun. 1979, 87, 693.
35. Li, C. H. in "Growth Hormone and Other Biologically Active Peptides"; Pecile, A.; Müller, M. M. (eds.) Excerpta Medica Int. Congr. Ser. 1980, 145, 45.
36. Mains, R. E.; Eipper, B. A.; Ling, N. Proc. Natl. Acad. Sci. USA. 1977, 74, 3014.
37. Roberts, J. L.; Herbert, E. Proc. Natl. Acad. Sci. USA. 1977, 74, 4826, 5300.
38. Herbert, E.; Roberts, J.; Phillips, M.; Allen, R.; Hinman, M.; Budorf, M.; Policastro, P.; Rosa, P. Front. Neuroendocrinol. 1980, 6, 67.
39. Nakanishi, S.; Inoue, A.; Kita, T.; Nakamura, M.; Chang, A.C.Y.; Cohen, S. N.; Numa, S. Nature. 1979, 278, 423.
40. Bewley, T. A.; Li, C. H. Adv. Enzymol. 1975, 42, 73.
41. Chrambach, A.; Yadley, R. A.; Ben-David, M. Endocrinology. 1973, 93, 848.
42. Mills, J. B.; Reagan, C. R.; Rudman, D. J. Clin. Invest. 1973, 52, 2941.
43. Li, C. H.; Gráf, L. Proc. Natl. Acad. Sci. USA. 1974, 71, 1197.
44. Li, C. H.; Bewley, T. A. Proc. Natl. Acad. Sci. USA. 1976, 73, 1476.
45. Bewley, T. A.; Li, C. H. Biochemistry. 1978, 17, 3315.
46. Li, C. H.; Bewley, T. A.; Blake, J.; Hayashida, T. Proc. Natl. Acad. Sci. USA. 1977, 74, 576.
47. Li, C. H.; Blake, J.; Hayashida, T. Biochem. Biophys. Res. Commun. 1978, 82, 217.
48. Li, C. H.; Blake, J. Proc. Natl. Acad. Sci. USA. 1978, 76, 6124.
49. Li, C. H. in "Hormonal Proteins and Peptides 8" Li, C. H. (ed.); Academic Press: New York, 1980, p. 2.
50. Birk, Y.; Li, C. H. Proc. Natl. Acad. Sci. USA. 1978, 75, 2155.
51. Li, C. H.; Houghten, R. Int. J. Peptide Protein Res. 1978, 12, 114.

52. Li, C. H. Proc. Natl. Acad. Sci. USA. 1978, 75, 1700.
53. Burstein, S.; Grumbach, M. M.; Kaplan, S. L.; Li, C. H. Proc. Natl. Acad. Sci. USA. 1978, 75, 5391.
54. Ramachandran, J.; Sairam, M. R. Arch. Biochem. Biophys. 1975, 167, 294.
55. Aggarwal, B. B.; Papkoff, H. Biochem. Biophys. Res. Commun. 1979, 89, 169.
56. Ward, D. N. in "Structure and Function of the Gonadotropins"; McKerns, K. W. (ed.) Plenum Press: New York, 1978, p. 31.
57. Moyle, W. R.; Bahl, O. P.; Marz, L. J. Biol. Chem. 1975, 250, 9163.
58. Sairam, M. R.; Schiller, P. W. Arch. Biochem. Biophys. 1979, 197, 294.

RECEIVED October 1, 1981.

Covalent Binding of Trypsin to Hydrogels

WILLIAM H. DALY and F. SHIH[1]
Louisiana State University, Department of Chemistry, Baton Rouge, LA 70803

Covalent binding of trypsin to hydrogels, i.e., derivatives of poly(2-hydroxyethyl methacrylate), PHEMA has been effected. Reaction of p-nitrobenzenesulfonyl isocyanate with PHEMA or copolymers containing HEMA and varying amounts of styrene followed by reduction of the nitro group with sodium dithionite produced p-aminobenzenesulfonyl carbamates of hydrogels (p-ABSC-HEMA). The precise composition of the hydrogel carbamates was determined by nonaqueous titration of the carbamate moiety with tetrabutylammonium hydroxide in isopropanol/dimethylformamide.

Treatment of p-ABSC-HEMA with glutaraldehyde produced enzyme supports capable of binding up to 55 wt % trypsin. Incorporation of hydrophobic styrene units into the support reduced the capacity to 2-4 wt % but enhanced the specific activity of the trypsin. The esterase activity of bound trypsin, assayed with TAME, was found to range from 11% to 45% of that exhibited by the free trypsin. Active-site titration of a PHEMA-trypsin conjugate with p-nitrophenyl-p'-guanadinobenzoate HCl indicated the active species to be 31% of the total amount of protein bound. Thus, 45% of the active sites present in free trypsin survived the immobilization process.

In recent years, a growing interest in hydrogels as solid supports for enzyme, hormones and pharmacons has become evident. Poly(hydroxyethyl) methacrylate), PHEMA, has been utilized to entrap trypsin and glucose oxidase (1) but the radical systems

---

[1] Current address: United States Department of Agriculture, Agricultural Research Service, Southern Regional Research Center, New Orleans, LA 70179.

employed deactivated a significant fraction of the enzymes.
Higher retention of enzymatic activity was observed when entrapment was effected using low temperature, radiation induced polymerization of enzyme-monomer mixtures in a glassy state (2).
Covalent binding of trypsin to crosslinked acrylamide-HEMA copolymers has been achieved by activating the support with cyanogen bromide; the immobilized enzyme exhibited 35% of the initial free enzyme activity (3). Hydrogels containing PHEMA and varying other components proved to be non-thrombogenic and stabilizing supports for a number of enzymes (4). A combination of entrapment and covalent binding has been particularly effective; copolymerization of acrylamide and acryloxysuccinimide to an active prepolymer followed by crosslinking with diamines in the presence of enzymes produced active enzyme matrixes from more than sixty enzymes (5). Copolymers of HEMA with p-acetaminophenylethoxy methacrylate or p-nitrophenol esters of ω-carboxymethacrylamide have been employed successfully to covalently bind enzymes (6).

Although the cyanogen bromide technique for activating hydroxyl containing polymeric supports has gained widespread acceptance (6), problems with removal of residual cyanide ions (7) has prompted a search for alternate activation techniques. Recently, the application of 2-(3-aminophenyl)-1,3-dioxolane as a synthon for diazotized or isothiocyanate functionalized polyvinyl alcohol has been described (8). We have elected to use p-nitrobenzenesulfonyl isocyanate as a synthon with general utility for the introduction of active binding sites on hydroxyl containing polymers. Using hydrogels, we have demonstrated the utility of this binding agent by immobilizing trypsin.

Experimental

Reagents. Hydroxyethyl methacrylate (HEMA) was extracted with hexane to remove bis-esters and distilled in vacuo, b.p. 69°/0.1 mm Hg. p-Nitrobenzenesulfonyl isocyanate (b.p. 160°/4 mm Hg) was synthesized in 63% yield by phosgenation of p-nitrobenzenesulfonamide in the presence of butyl isocyanate (9). Homo- and copolymers of HEMA were prepared by solution polymerization in DMF in the presence of benzoyl peroxide and the results are summarized in Table I. The hydrolytic stability of the hydrogels was estimated by slurrying 0.5-1.0 g of polymer in 5.0 mL 0.35 N KOH in 10 mL culture tubes equipped with teflon lined screw caps. The samples were heated at 100° for up to 24 hr in an aluminum constant temperature block. Immediately upon removal from the heating block, the samples were cooled and acidified with 6N HCl. The precipitated gel was washed and soaked in distilled water for 30 min before the carboxylic acid content was estimated by titration with standard NaOH.

Preparation of Hydrogel p-Nitrophenyl Carbamates. A 20 mL solution of hydrogel (2-3.0 g) in pyridine was treated with 3 g

TABLE I. SYNTHESIS OF HYDROGELS

Hydrogel System	Monomer Feed[a] HEMA g	Styrene g	Monomer Ratio in Copolymer[b] HEMA:STY	Precipitant	Conv. %	Mol. Wt.[c] 10^4	Solvents
Poly(2-Hydroxyethyl Methacrylate) PHEMA (1a)	18			Benzene	79	3.1	DMF Pyridine
Copoly(2-Hydroxyethyl Methacrylate-Styrene), CP(HEMA-STY) (2a)	13	10	5 : 4	Water	80	2.7	THF, DMF Pyridine
Copoly(2-Hydroxyethyl Methacrylate-Styrene), CP(HEMA-STY) (3a)	5	19	1 : 5	Water	78	2.6	Benzene CHCl3, THF DMF, Pyridine
Poly(Hydroxypropyl Acrylate) PHPA (4a)	HPA, 30 g			Benzene	70	2.4	DMF Pyridine
Poly(2-Hydroxyethyl Acrylate) PHEA (5a)	HEA, 30 g			Benzene	60	2.2	DMF Pyridine

[a] Monomer:solvent (DMF) - 1:5 v/v, initiator = BPO (0.3 wt % of solvent), reaction time = 14 hours, temperature = 70°C. [b] Based on oxygen content in the copolymer. [c] GPC (polystyrene standard)

of p-nitrophenyl isocyanate. The solution was heated at 70° for
6 hr before precipitating the polymer adduct by adding the reaction solution to 200 mL of water. The adduct was redissolved in
THF, reprecipitated in isopropanol and dried in vacuo at 60°
overnight before the extent of carbamation was estimated by elemental analysis (Table II).

Preparation of Hydrogel Arylsulfonyl Carbamates. To a solution of hydrogel (1-4.0 g) in dry pyridine (20-40 mL) was added
either p-toluenesulfonyl isocyanate (pTSI), 3.0 g, or p-nitrobenzenesulfonyl isocyanate (p-NBSI), 0.5-3.0 g and the mixture
was heated at 70° under nitrogen for 4 hr. The adduct was isolated by precipitation in benzene or ethanol. Reprecipitation
from acetone or DMF yielded the desired products, IR (film) 3170
cm^{-1} (-NH), 1530 cm^{-1} (-NO_2), 1340, 1180 cm^{-1} (-SO_2). The precise degree of substitution was determined by non-aqueous
titration (Table II).

Reduction of Hydrogel p-Nitrobenzenesulfonyl Carbamates. A
solution of p-NBS carbamate, 1.0 g, in 20 mL of THF was blended
with 15 mL of 10% aqueous $Na_2S_2O_4$. The two phase system was
stirred vigorously and heated to gentle reflux for 4-6 hr. The
reduction was monitored by observing the disappearance of nitro
absorbance at 1530 cm^{-1}. Evaporation of the THF phase, followed
by exhaustive washing with distilled water yielded the reduced
adduct. Loss of product during the isolation procedure due to
partial solubility in water was normally observed. The extent of
reduction is reported in Table II.

Non-Aqueous Potentiometric Titration of Hydrogel Arylsulfonyl
Carbamates. An approximately 0.1 N solution of tetrabutylammonium hydroxide (TBAH) in isopropanol was prepared by passing
100 mL of a saturated isopropanol solution of tetrabutylammonium
iodide through an Amberlite IRA-400 column which has been converted to the hydroxide form. The column was washed with 400 mL
of isopropanol and the eluates were combined and standardized by
titrating benzoic acid dissolved in DMF.

A sample of the hydrogel carbamate (50-80 mg) was dissolved
(or slurried) in 3 mL DMF. The solution was titrated in a microtitration vessel equipped with a 5.5 mm combination electrode.
The end points and half neutralization potentials (HNP) were
identified by the inflections in the EMF (MV) titration curve. A
0.3 thymol blue isopropanol solution also served as an effective
end point indicator (yellow to blue).

Glutaraldehyde Immobilization of Trypsin. A sample of hydrogel-p-aminobenzenesulfonyl carbamate (0.1-0.2 g) was suspended in
5.0 mL pH 7 phosphate buffer solution in a 10 mL vial equipped
with a stirring bar and Teflon lined screw cap. To the suspension was added an excess of 50% glutaraldehyde (0.5 mL), and the

mixture was stirred at room temperature for 1.5 hours. After centrifuging, the supernatant was removed by decantation, the residue was washed with five 7.0 mL portions of pH 7 buffer solution, and the washings were discarded. To the washed residue was added 8.0 mL pH 7 buffer and an excess of trypsin (20-50 mg). The mixture was stirred at 5°C overnight (17-20 hours). The residue was then washed with aliquots of pH 7 buffer until the wash exhibited no activity toward TAME. The immobilized enzyme was stored in pH 8 borate buffer (10-30 mL) at 4°C.

Titrimetric Analysis of Trypsin Activity. Into the microtitration vessel were introduced 0.6 mL borate pH 8 buffer, 0.2 mL 0.01 M $CaCl_2$ solution, and various increments of TAME, i.e., 0.1 mL, 0.2 mL, 0.4 mL, 0.7 mL, and 1.4 mL of a freshly prepared 0.08 M TAME solution in pH 8 buffer. At zero time, an aliquot of 0.1 mL of the immobilized trypsin stock suspension or 0.1 mL of a known free trypsin solution was introduced, and 0.1 N NaOH solution was added dropwise to maintain a constant pH reading. The time and volume of base added were recorded.

Results and Discussion

Arylsulfonyl isocyanates are very reactive acylating agents which form adducts with alcohols, amines, polysubstituted double bond and activated aromatic rings (9). The arylsulfonyl carbamates produced upon addition to alcohols exhibit an acidity comparable to that of benzoic acid. Reaction of cellulose with p-toluenesulfonyl isocyanate, which produces the corresponding carbamate without pretreatment, has been utilized to produce a cation exchange resin (10). In neutral and alkaline media, the ionized arylsulfonyl carbamates enhance the hydrophilicity of hydrophobic copolymers and would provide counterions for trace metals frequently associated with enzymes. Thus we elected to design a coupling agent based upon arylsulfonyl isocyanate chemistry to take advantage of the diversified chemistry exhibited by these reagents. p-Nitrobenzenesulfonyl isocyanate appeared to be the best candidate for evaluation.

Although hydrogels exhibit excellent compatability with enzymes, the gels generally exhibit poor mechanical properties. Tear resistance is very low; data on PHEMA indicates a tensile strength of 50 psi, an ultimate elongation of 70% and an elastic molulus of 140 psi (11). Since copolymerization of HEMA is readily effected (12), we decided to strengthen the hydrogel and moderate the hydrophilicity of the support by incorporation of styrene. Soluble copolymers were prepared to facilitate characterization but the corresponding compositions could be prepared in a crosslinked form by incorporating divinylbenzene. The efficacy in enzyme binding by the copolymer supports was compared with that of PHEMA using trypsin as a standard. The low hydrolytic stability of hydroxyalkyl acrylates forced us to limit our studies to methacrylate derivatives (Figure 1).

Hydrogel Modification. Treatment of either pyridine or tetrahydrofuran solutions of the hydrogels with p-nitrobenzenesulfonyl isocyanate (PNBSI) at 70° afforded the corresponding arylsulfonyl carbamates in 40-100% yield. The extent of substitution increased when styrene copolymers were treated indicating that the hydrophobic character of the support favors modification. The reaction of PHEMA with p-toluenesulfonyl isocyanate (PTSI) and p-nitrophenyl isocyanate (PNPI), a common enzyme coupling agent, was conducted under similar conditions and the ease of carbamation decreased in the order: PNBSI > PTSI > PNPI. As expected, the carbamation enhanced the hydrophilicity of the gels; the adducts were more swollen in water and tended to dissolve in base. The stability of the hydrogel adducts in acid or base appeared to parallel that of the original gels. For example, when PHEMA-PNBS carbamate was subjected to 2 $\underline{N}$ HCl or 2 $\underline{N}$ NaOH for 12 hours at room temperature, the extent of substitutent loss was found to be 4% and 17%, respectively, based upon loss of nitrogen content.

To activate the hydrogel adducts for enzyme immobilization, it was necessary to reduce the nitro function to the corresponding amine. From the many techniques available to effect this modification, we selected sodium dithionite as the mildest procedure with the least potential problems with residual reagent. Since the hydrogels were quite hydrophilic it was possible to use a two phase solvent system (THF/H_2O) to effect the reduction. The arylsulfonyl carbamates were reduced selectively under these conditions; no reduction of p-nitrophenyl carbamate was observed.

TABLE II. MODIFICATION OF HYDROGELS

Hydrogel	Derivative PTS(b) Carbamate % substn.	PNBS(c) Carbamate % substn.	PABS(d) Carbamate % substn.	PNP(e) Carbamate % substn.	PAP(f) Carbamate % substn.
PHEMA (1a)	1b, 28	1c, 43	1d, 65	1e, 18	∿0
CP(HEMA STY) (2a)	2b, 30	2c, 60	2d, 84	2e, 25	∿0
CP(HEMA-STY) (3a)	3b, 40	3c, 100	3d, 100	3e, 34	∿0
PHPA (4a)	4b, 60	4c, 66	4d, 70	—	—
PHEA (5a)	5b, 37	5c, 51	5d, 65	—	—

b) p-Toluenesulfonyl; c) p-Nitrobenzenesulfonyl; d) p-Aminobenzenesulfonyl carbamate; e) p-Nitrophenyl; f) p-Aminophenyl

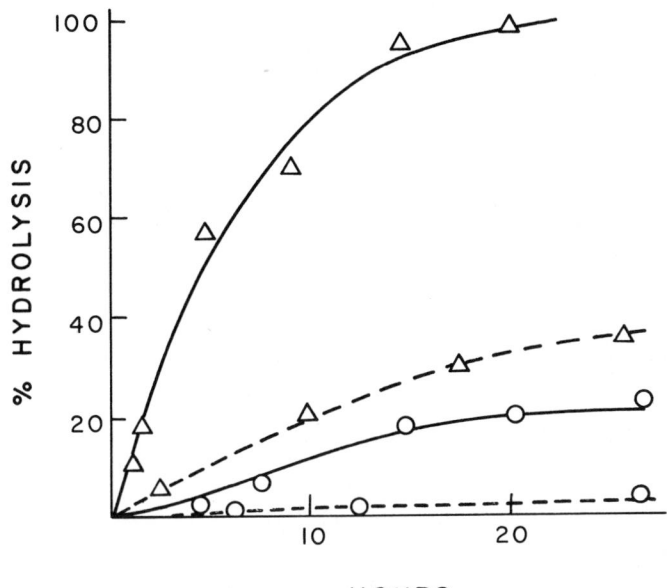

Figure 1. Hydrolysis of poly(2-hydroxyethyl methacrylate) (○) and poly(hydroxypropyl acrylate) (△) in 0.4 M KOH. Conditions: —— 100°; and – – – 25°.

A major problem encountered in polymer modification is the precise analysis of the substituents introduced. Multiple modifications compound this problem and introduce further sources of error. However, we observed that the relative acidity of the arylsulfonyl carbamates was dependent upon the substituents and that the various end points could be resolved using nonaqueous titration techniques. Using dimethylformamide as a solvent, and tetrabutylammonium hydroxide (TBAH) in isopropanol as a titrant, the potentiometric titrations can be conducted with normal glass and calomel electrodes. End points were recognized at the potentiometric breaks or with the aid of indicators and the relative strengths of the acid could be expressed as half neutralization potentials. The hydrogel derivatives exhibit half neutralization potentials within ± 200 mv, which corresponds roughly to pKa values between 5 and 9 in aqueous media (Scheme I). Combining elemental analysis with nonaqueous titration enables us to determine the precise composition of a hydrogel derivative.

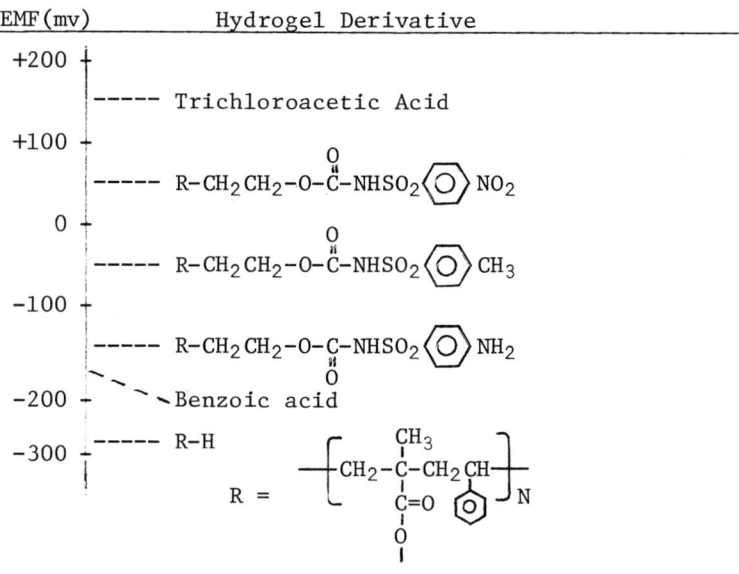

Scheme I. HALF NEUTRALIZATION POTENTIALS IN DMF

Trypsin Immobilization. The acidity of the arylsulfonyl carbamates limited the methods for immobilizing trypsin. Although treatment of a p-aminobenzenesulfonyl carbamate adduct with isoamyl nitrite generated a diazonium salt, no enzyme coupling could be detected. Thus, we were forced to couple trypsin to the hydrogel adducts with glutaraldehyde (Scheme II). Control runs demonstrated that simple adsorption was not responsible for the observed activity. The esterase activity was ascertained utilizing α-N-p-toluenesulfonyl-L-arginine methyl ester (TAME) as a substrate and monitoring extent of reaction by titration. The total enzyme bound was estimated by complete hydrolysis of the enzyme conjugate and determination of selected amino acids by gas chromatography (13). The amino acids were double derivatized; first esterification with n-propanol and then treatment with heptafluorobutyric anhydride to acylate the amino groups. The quantitative GC analysis was simplified by using a difunctional standard, DL-α-ε-diaminopimelic acid. Figure 2 illustrates a typical chromatogram obtained from hydrolysis of enzyme conjugate 3. For comparison, the chromatogram of the free trypsin, used as the reference standard, is reproduced in Figure 3. By selecting four or five well resolved peaks and ratioing them to the corresponding peak areas of pure trypsin, one obtains a very reliable estimate of the protein bound (Table III).

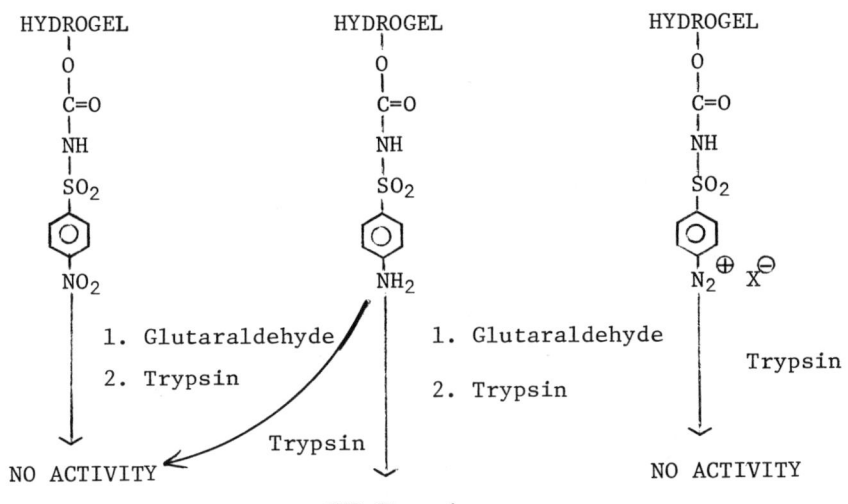

Scheme II. TRYPSIN IMMOBILIZATION

Active-site Titration. In investigating immobilized enzymes, it is important to be aware of another highly significant consideration, i.e., the relative amounts of active, partially active, and inactive enzyme molecules per total bound enzyme. In principle, this information can be obtained with active-site directed titration reagents. For example, Fritz et al. (14) used p-nitrophenyl-p'-guanidinobenzoate (NPGB) to determine the "active" trypsin molecules in a trypsin cellulose conjugate. NPGB proved to be an excellent active-site titrant for tyrpsin; Ford and co-workers (15) used it on trypsin covalently linked to porous glass particles, employing a recirculating reactor system capable of providing accurate results at active enzyme concentration of

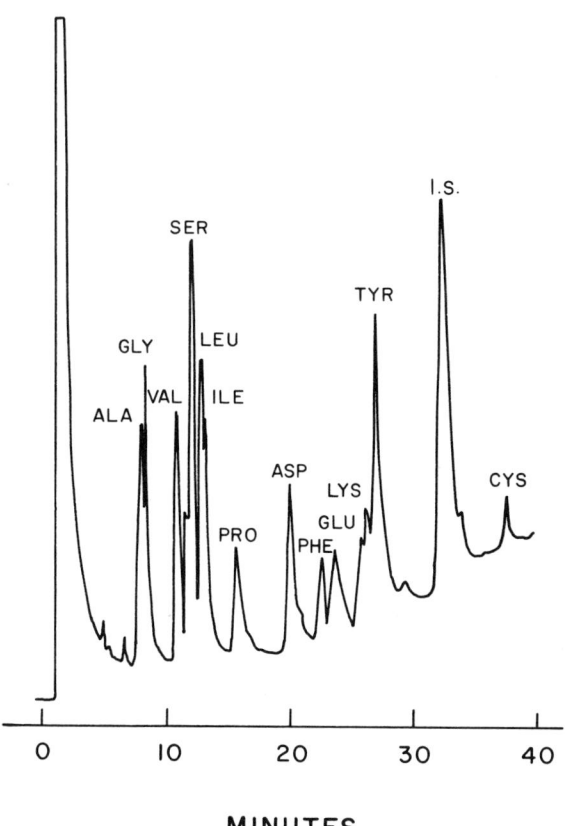

Figure 2. Gas chromatogram of the N-HFB n-propyl derivatives of the hydrolysate of trypsin CVB hydrogel sulfonyl carbamate (3f). Conditions: 12 ft. × 1/4 in. glass column packed with 3% OV-1 on Chromasorb-W; initial temp was 100° for 4 min, programed to 250° at 4°/min.

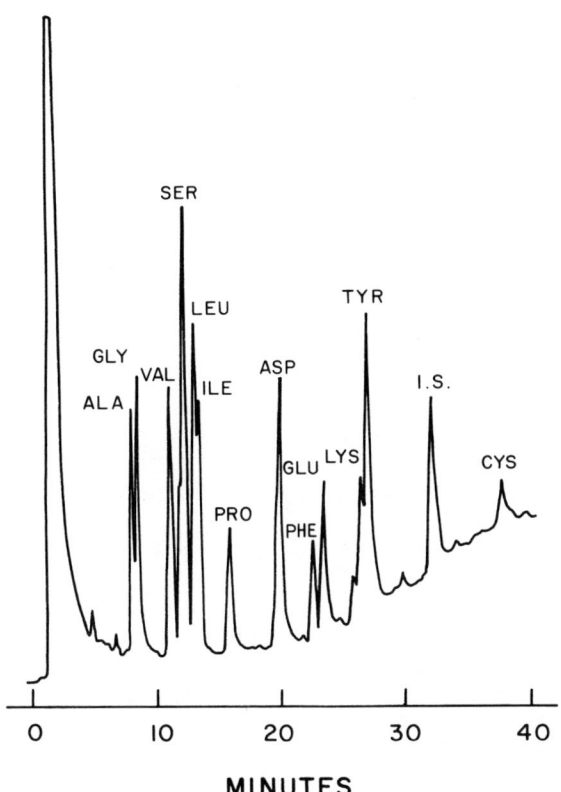

Figure 3. Gas chromatogram of the N-HFB n-propyl derivatives of the hydrolysate of free trypsin. Conditions: 12 ft. × 1/4 in. glass column packed with 3% OV-1 on Chromasorb-W; initial temp. was 100° for 4 min, programed to 250° at 4°/min.

TABLE III. SUMMARY OF ACTIVITY ANALYSIS FOR TRYPSIN CVB HYDROGEL SULFONYL CARBAMATES

Analysis \ Trypsin-$_{cvb}$-Hydrogel-PABSC	PHEMA 1f	PHEMA 1g	CP(HEMA-STY) (1:1) 2f	CP(HEMA-STY) (1:1) 2g	CP(HEMA-STY) (1:5) 3f
Percent Carbamation	43.0	7.9	60.0	3.5	100
$\frac{\text{Mg Trypsin}}{\text{mL stock suspn.}}$	0.33	0.10	0.30	0.06	0.26
Enzyme bound wt percent	55.0	1.0	2.6	0.4	4.4
$K_m(app) \times 10^3 M$	6.8	18.0	8.2	13.0	16.0
$V'_{max} \frac{\mu \text{ mole}}{\text{Min-Mg}}$ a	44.0	60.0	30.0	125	47.0
Percent Sp. Activity	17.0	21.0	11.0	45.0	38.0

a) Free Trypsin: V'_{max} = 280 μ mole/Min-Mg

$K_m = 4.8 \cdot 10^{-3}$ M

1×10^{-6} moles trypsin per liter fluid volume. To demonstrate the feasibility of using the Ford method to determine the active-site of our immobilized enzyme systems, trypsin CVB-PHEMA-PABS-carbamate was treated in a circulation reactor with NPGB and the titration is illustrated in Figure 4. The amount of p-nitrophenol produced by the burst is equal to the amount of the active immobilized trypsin which, for this particular system, turns out to be 31% of the total bound enzyme. Active-site titrations of soluble trypsin were performed according to Chase and Shaw (16), and the active molecules for free trypsin was found to be 70% of the total protein involved. Consequently, the retention of active molecules for the immobilized enzyme was calculated 45%. The specific activity is 17% (Table III) for the same system so the efficiency of the system, based on the actually available active sites, was 38%. Thus, 62% of the initially active trypsin bound has lost its activity upon binding.

The use of kinetic assay methods requires several assumptions regarding the kinetic parameters of the immobilized enzyme and the influence of diffusion on the results. As has been discussed earlier, for many substrate-immobilized enzyme combina-

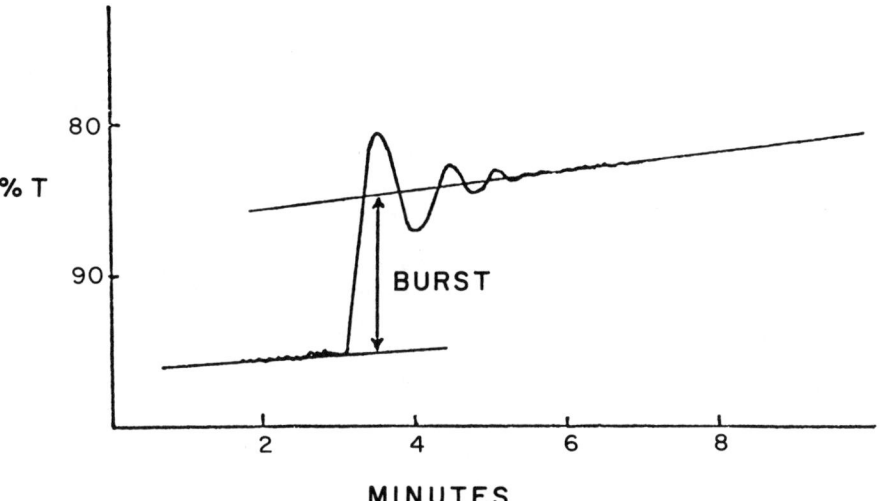

Figure 4. Active site titration recorder trace of 17.0 mg trypsin CVB hydrogel sulfonyl carbamate (1f) being titrated by 19.75 mL of 0.12 mM NPGB in pH 8.3 veronal buffer; burst, 0.060 A.

tions the rapid rate of reaction provides sizable concentration gradients both outside the immobilizing matrix and within its pores. These concentration gradients greatly retard the rate of reaction, even when the substrate concentration within the pores is much greater than the immobilized enzyme Km (17). This diffusional effect serves to decrease the apparent active enzyme concentration. In order to obtain valid active enzyme concentrations from a kinetic assay, the K_{cat} for the immobilized enzyme must be known. However, the effects of the micro-environment and the immobilization process itself cause the kinetic parameters of the immobilized enzyme to differ frequently from those of the soluble enzyme. This shift in K_{cat} of the immobilized enzyme as result of changes in the micro-environmental pH (pH optima shift) has been well documented (18).

Influence of Support Structure upon Trypsin Activity. The apparent Michaelis constant, Km(app), is much greater for immobilized enzymes than Km of free enzyme. In our systems Km (app) increases as the hydrophobic component increases (Table III), apparently substantiating the theory that Km(app) of an immobilized enzyme will be decreased by attachment to a hydrophilic polymer. Another trend shown in Table III is the increase of Km values with the decrease of the degree of carbamation. For example, in PHEMA, the value of Km is 6.8×10^{-3} M when its carbamation is 43%, but it jumps to 18×10^{-3} M with a much lower degree of substitution. The change is understandable when one realizes that with a higher percentage of carbamation the degree of ionization due to the acidic carbamate groups is greater and, at the same time, the enzyme to carrier ratio increases. The former factor enhances the hydrophilicity of the system, and the latter reduces the ratio of components that are foreign to the enzyme; both contribute to a smaller diffusion restriction and, as a result, a smaller Km(app) for the system with a higher degree of carbamation.

Table III also shows that trypsin-cvb-hydrogel derivatives retain in the range of 11-45% the original enzyme activity, which is about average for the covalently bound trypsin on polymers containing hydrogel. For comparison, Mosbach (3) immobilized trypsin on a crosslinked copolymer of acrylamide and hydroxyethylmethacrylate via CNBr coupling, and analyzed it photometrically with α-N-benzoyl-Dl-arginine-p-nitroanilide (BAPNA), reporting a 35% retention of activity. O'Driscoll and co-workers (1) however, reported an equivalent of 1.3% efficiency when trypsin was physically entrapped in a crosslinked PHEMA gel and assayed with TAME.

Some reports in literature have shown that the specific activity of a covalently bonded enzyme increases as the solubility of its support increases (19-20). Others have observed that a rough inverse correlation exists between the activity of an immobilized enzyme and the amount of protein bound on the support

(21). These relationships were found true to some extent in our experiments. As seen in Table III, the bound enzyme efficiency increases when the amount of protein immobilized decreases sixfold, and it also increases with a substantial decrease of enzyme content. However, it should be noted that, in both cases, the increase in efficiency occurs on polymer supports with a supposedly offsetting effect of decreasing hydrophilicity. Thus, regardless of the effect of its hydrophilicity, a hydrogel-enzyme conjugate appears to exhibit a higher degree of activity retention with a lower enzyme load. An exception is in the case of (2f) and (3f), in which a relatively small increase of enzyme bound results in a substantial increase of activity. In addition, this increase in efficiency is noticeably accompanied by a decrease of hydrophilicity of the polymer support. The fact that hydrogel systems with high contents of styrene or low degrees of carbamation seem most often associated with high values of specific activity leads us to believe that phase separation in copolymers containing fractions of highly different hydrophilicity and interactions between enzyme molecules and hydrogel (or carbamates) are probably the causes for the discrepancies. Thus, when the styrene content in a hydrogel system is high or the percent carbamation is low, phase separation occurs; the hydrophobic section effectively forces the smaller populations of the hydrophilic carbamate-trypsin conjugates to be separated, localized, and concentrated. When this happens the enzyme molecules are in a more natural environment, and they exhibit better activity efficiency, such as in the cases of (2g) and (3f). On the other hand, when the contents of the hydrogel monomers or carbamate substituents in the systems are high, there is a possibility that some interpenetration between the bound enzyme and the hydrogel network might occur. The enzyme molecules are in a less desirable environment compared with systems with higher degrees of phase separation, and their structure might even suffer some disruption. As a result, the favorable diffusion effect due to the presence of highly hydrophilic elements might still enhance the activity of the bound enzyme but not to the extent as expected. This may explain in large part the relatively low values of specific activity exhibited by (1f), (1g) and (2f). However, it is extremely difficult to ascribe precisely the cause and magnitude of an alternation in the activity of an immobilized enzyme.

Literature Cited

1. O'Driscoll, K. F.; Izu, M.; Konus, R. Biotech. Bioeng. 1972, 14, 847; O'Driscoll, K. F.; Kapoulas, A.; Albisser, A. M.; Gander, R. Polymer Preprints 1975, 16, 372.
2. Yoshida, M.; Kumakura, M.; Kaetsu, I. J. Macromol. Sci.-Chem. 1980, A14, 541, 555.
3. Mosbach, K. Acta Chem. Scand. 1970, 24, 2084.

4. Venkataraman, S.; Horbett, T. A.; Hoffman, A. S. Polymer Preprints 1975, 16, 197. Ratner, B. D.; Horbett, T. A.; Hoffman, A. S.; Hauschka, S. D. J. Biomed. Mater. Res. 1975, 9, 407.
5. Pollak, A.; Blumenfeld, H.; Wax, M.; Baughn, R. L.; Whitesides, G. M. J. Am. Chem. Soc. 1980, 102, 6324.
6. Turkova, J. "Methods in Enzymology", 1976, 44, 66.
7. Pommerening, K.; Jung, W.; Kühn, M.; Mohr, P. J. Poly'm. Sci., Polym. Sym. 1979, 66, 185.
8. Manecke, G.; Vogt, H. G. Angew. Makromol. Chem. 1979, 78, 21.
9. Ulrich, H.; Sayigh, A. A. R. Angew. Chem. Internal. Bd. 1966, 5, 704.
10. Rousseau, R. W.; Callihan, C. D.; Daly, W. H. Macromolecules 1969, 2, 502.
11. Gregonis, D. E.; Chen, C. M.; Andrade, J. D. Polymer Preprints 1975, 16, 349.
12. Wiley, R. H.; Sale, S. S. J. Polymer Sci. 1960, 42, 491.
13. Gehrke, C. W.; Roach, D.; Zumwalt, R. W.; Stalling, D. L.; Wall, L. L. "Quantitative Gas-Liquid Chromatography of Amino Acids in Protein and Biological Systems", Analytical Biochemistry Laboratories, Inc., Columbia, Mo., 1968.
14. Fritz, H.; Gebhardt, M.; Molster, R.; Illchmann, K.; Hochstrasser, K. Z. Physiol. Chem. 1970, 351, 571.
15. Ford, J. R.; Chambers, R. P.; Cohen, W. Bioch. et. Biophys. Acta 1973, 309, 175.
16. Chase, Jr., T.; Shaw, E. Biochem. Biophys. Res. Commun. 1961, 29, 508.
17. Lambert, A. H.; Ford, J. R.; Cohen, W.; Chamber, R. P. "The Influence of Diffusion on the Kinetics Behavior of Enzymes Immobilized in Porous Solids", presented at San Francisco AICHE Meeting, Dec. 1971.
18. Katchalski, E.; Silman, I.; Goldman, R. Adv. Enzymol. 1971, 34, 445.
19. Axen, R.; Myrin, P. A.; Janson, J. C. Biopolymers 1970, 9, 401.
20. O'Neill, S. P.; Wykes, J. R.; Dunnill, P.; Lilly, M. D. Biotechnol. Bioeng. 1971, 13, 319.
21. Axen, R.; Ernback, S. Eur. J. Biochem. 1971, 18, 351.

RECEIVED July 7, 1981.

Chitin—Protein Complexes

Ordered Biopolymer Composites

JOHN BLACKWELL, LOUIS T. GERMINARIO[1], and MARK A. WEIH[2]
Case Western Reserve University, Department of Macromolecular Science, Cleveland, OH 44106

The skeletal materials of insects and many other "lower" animals are biopolymer composites, in which chitin (polysaccharide) fibrils are separated by a matrix of protein. X-ray fiber diagrams show that in many of these systems both the fibrous and matrix components are ordered. We have developed a model for the three-dimensional structure of a typical insect chitin-protein complex, that from the ovipositor of the ichneumon fly *Megarhyssa*. The x-ray pattern shows that crystalline α-chitin fibrils are arranged hexagonally with a center to center distance of 7·25nm. This hexagonal packing is confirmed by electron microscopy of stained ovipositor sections. The protein forms a helical sheath around each chitin fibril, with six protein subunits repeating in 3·06nm along the fiber axis. Fourier reconstruction of the electron micrographs also reveals a sixfold arrangement of subunits in the protein matrix. The structure provides a basis for the discussion of fiber-matrix interactions in these and other structural tissues, such as collagen and cellulose containing systems.

The polysaccharide chitin forms the component of the skeletal materials of many lower animals, (<u>1</u>, <u>2</u>, <u>3</u>) and serves in an analogous way to collagen in mammalian connective tissues and cellulose in plant cell walls. All of these tissues can be classified as fiber-matrix composites: in the cell walls of higher plants the cellulose fibrils are surrounded by other polysaccharides, proteins, and also lignin in some cases; in

---

[1] Current address: Tennessee Eastman Company, Kingsport, TN 37662.
[2] Current address: Lord Corporation, Erie, PA 16512.

mammalian connective tissues collagen fibrils are surrounded by a gel of proteoglycans and other proteins; and in chitinous tissues the fibrils are in a matrix of protein. In addition, the animal tissues are frequently calcified. The mechanical properties of these skeletal materials necessarily depend on the structures of the fiber and matrix phases and their mutual interaction. At present we have extensive knowledge of the physical structures of all three types of fibers, but information with regard to the structure of the matrix has been more difficult to obtain. For the cellulose and collagen containing tissues, there has been extensive study of the matrix chemistry, but x-ray studies indicate that the physical structures are amorphous. The chitinous tissues have received less attention, but it has been shown that in certain of these systems the matrix proteins are ordered, so that it is possible to develop a three-dimensional model for the fiber-matrix structure and interactions. The present paper describes our studies using x-ray diffraction and electron microscopy to derive the structure of one of the more ordered chitin-protein complexes, that from ovipositor of the ichneumon fly *Megarhyssa*.

Chitin is a repeating homopolymer of β(1,4)-linked N-acetyl-D-glucosamine (NAG) and is obtained after chemical or enzymic deproteinization of the native tissues. Such specimens of chitin generally contain some unacetylated residues, (4, 5) and there has been some speculation that these may be the points for covalent linkage to the protein. (6-9) Much further work is necessary in this area, but evidence has been presented for covalent chitin-protein linkages at some of the chitin chain ends in certain tissues. It is interesting to note that the only reported chitin that is pure poly NAG occurs in the spines of certain marine diatoms, and exists completely free of protein. (10) Chitin has been shown to exist naturally in two polymorphic forms, known as α- and β-chitins. (11) In both forms the chitin chains have a 2_1 helical conformation, (12, 13, 14) with two NAG residues repeating in approximately 1·03nm similar to that observed for cellulose. (15, 16) The two forms of chitin differ in terms of the polarity of adjacent chains, which are antiparallel in α- and parallel in β-chitin. This leads to different intermolecular hydrogen bonding networks, (17, 18) with a result that β-chitin is swollen by water and α-chitin is not, which may explain why α-chitin is by far the more common form found in nature. The chitin in *Megarhyssa* ovipositor is α-chitin, and has the structure shown in fig. 1, as was determined previously in this laboratory. (18)

Electron micrographs of insect cuticles in which the protein is stained e.g. with uranyl acetate show a hexagonal array of (unstained) chitin fibrils in a matrix of protein. Neville et al (19) have reported microfibrils with diameters of 2·4 - 3·0nm separated by 5·1 - 8·3nm in a survey of insect cuticles from a wide variety of species. Rudall (20) observed a hexagonal lattice dimension of a = 6·9nm for *Megarhyssa* ovipositor. He also

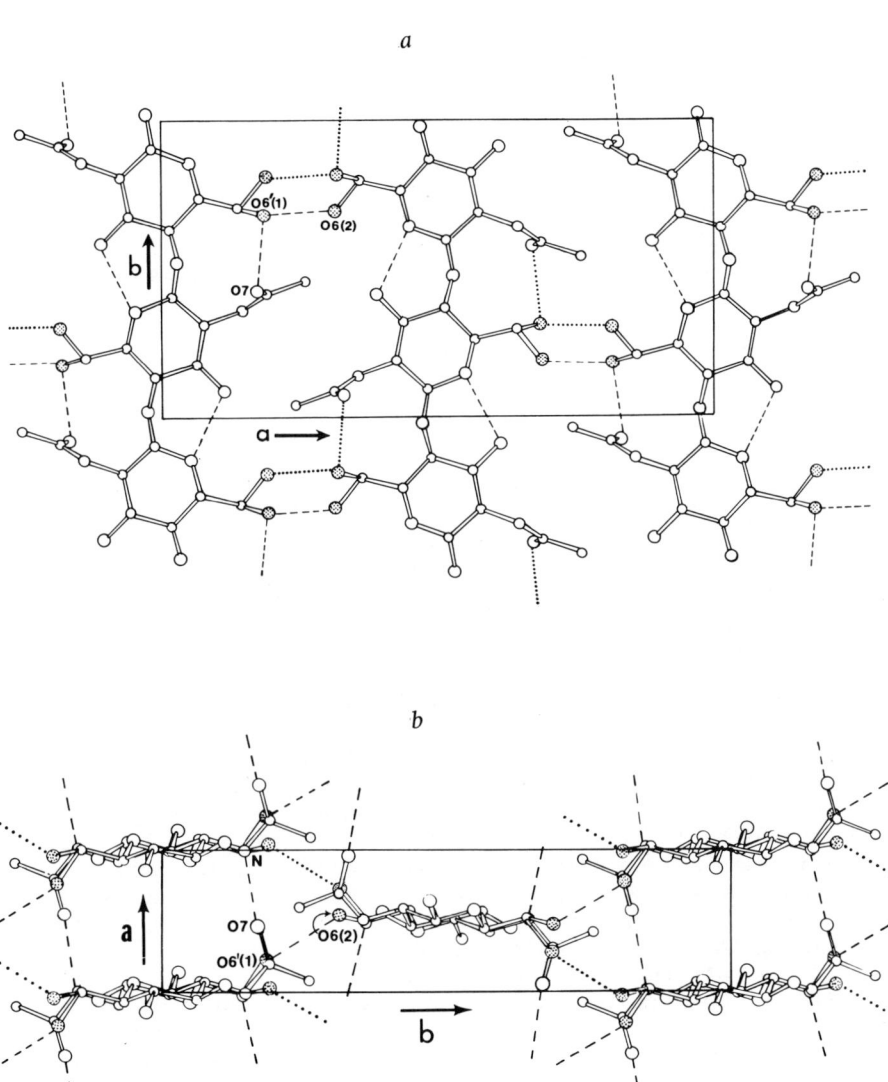

Figure 1. Structure of α-chitin. Key: a, ac projection and b, ab projection. (Reproduced, with permission, from Ref. 18. Copyright 1973, Academic Press.)

showed (20, 11) that this and many other insect cuticles gave x-ray patterns that indicated that the protein matrix is ordered. This led to a classification (21, 22) of the chitin-protein complexes based on their fiber repeats of 3·1, 4·1, or 6·2nm, which are respectively three, four, and six times the chitin repeat.

At present relatively little is known about the chemistry of the matrix protein, although significant differences have been reported between proteins extracted from different tissues. Extraction data indicate the presence of numerous protein moieties, (23, 24, 25, 6) but interpretation is difficult because of extensive cross-linking (tanning) in mature cuticles. Rudall and Kenchington (21) report that the evidence for ordering of the protein is eliminated by extraction with 7M urea, suggesting a non-covalent interaction between protein and chitin. Rudall (22) also cites Skerrow, who showed evidence for a basic protein subunit of molecular weight 27,000 extracted in 7M urea from young honeybee (*Apis mellifera*) cuticle, a tissue giving a similar x-ray pattern to that for *Megarhyssa* ovipositor.

In the work described below we have derived a structure for the chitin protein complex of *Megarhyssa* ovipositor, for which confirmatory evidence has been obtained by Fourier reconstruction of electron micrographs. Details of this work have been published separately (see refs. 26 and 27).

Materials and Methods

Mature female ichneumon flies (*Megarhyssa lunator* and *Megarhyssa atrata*) were generously provided by Dr. S. Teraguchi of this university, Dr. A.G. Nielson of Ohio Agriculture Research and Development Center, Wooster, Ohio, and Dr. K.M. Rudall, of Auckland, New Zealand. Fresh and museum specimens proved to be indistinguishable by x-ray and electron microscope techniques. The ovipositors were easily separated longitudinally into three sections (1st and 2nd volvifers) and cut into 1 cm lengths to serve as x-ray specimens. Deporteinized specimens were prepared by boiling in 5% KOH solution for 24 hrs.

X-ray patterns were recorded on Kodak No-Screen film using Ni filtered CuKα radiation from a Rigaku Denki rotating anode source. Small angle and wide angle patterns were obtained using a Rigaku Denki and Searle (toroidal) camera respectively. An E.D.P. scanning optical densitometer was used to measure intensities and to determine the small angle d-spacings and the positions of the maxima on the protein layer lines.

Specimens for electron microscopy were prepared as thin cross-sections and stained with uranyl acetate and lead citrate. A JEOL 100B electron microscope was used to obtain the electron micrographs. Reconstruction of sections of these micrographs by Fourier filtration methods was done in collaboration with Dr. J. Frank of New York State Department of Health, Albany, N.Y.

(as described in refs. 28 and 29). In addition, a number of individual fibrils were selected where the neighboring protein appeared to be well ordered. These fibrils were averaged after optimization of their rotational orientation.

Results and Discussion

Figures 2 and 3 show the x-ray fiber diagrams of intact and deproteinized *Megarhyssa* ovipositor. Figure 3 is characteristic of α-chitin and most of the features of this pattern can also be seen in fig. 2. Thus it can be concluded that the intact ovipositor contains crystalline α-chitin. The additional data in fig. 2 are quite extensive and indicate that there is an ordered superstructure. Firstly, in the small angle region there are three equatorial reflections at d = 6·29, 3·62, and ∿ 3·14nm. These can be indexed by a two-dimensional hexagonal lattice with dimension a = 7·25nm. This compares favorably with a ≃ 6·9nm reported previously by Rudall, (20) and a ≃ 7·5nm for the center to center distance of hexagonally packed chitin microfibrils seen in our electron micrographs (see below). Comparison of the squared Fourier transforms of solid cylinders with the observed small angle intensities (F^2) showed best agreement for a diameter of 7·25nm. We therefore envisage a structure comprised of microfibrils, each consisting of a crystalline core surrounded by a sheath of protein, which are closely packed on a hexagonal lattice as shown in fig. 4. The calculations also indicate that the protein and chitin have approximately the same electron density, and therefore the radius of the chitin core is not defined. The complex contains approximately 85% protein, from which a core diameter of 2·8nm can be calculated on volume considerations. This however assumes a homogenous distribution of the chitin and protein in the bulk structure. From our electron micrographs the fibril diameter is in the range 3 - 4nm and it is difficult to be more specific due to the granular appearance of the stained protein. It has been suggested by Atkins (30) that certain wide angle equatorial reflections due to the chitin may arise from a crystallite four chains wide, which would correspond to a dimension of 3·8nm. Thus the diameter of the chitin core is uncertain (within the above limits) but this is not critical to the analyses that follow for the arrangement of the protein subunits.

In the wide angle region of the x-ray pattern for the intact ovipositor there appear to be three layer lines for every one for the purified chitin: layer lines 3, 6, and 9 in fig. 2 are very similar to layer lines 1, 2, and 3 in fig. 3. The extra layer lines (1, 2, 4, 5, 7, 9, 10...) and the additional intensities on layers 3, 6, and 9 must arise from the protein (and its interaction with the chitin). The layer line repeat is 3·06nm, and thus we envisage a protein sheath structure with a repeat approximately three times that of the chitin, i.e. six NAG residues.

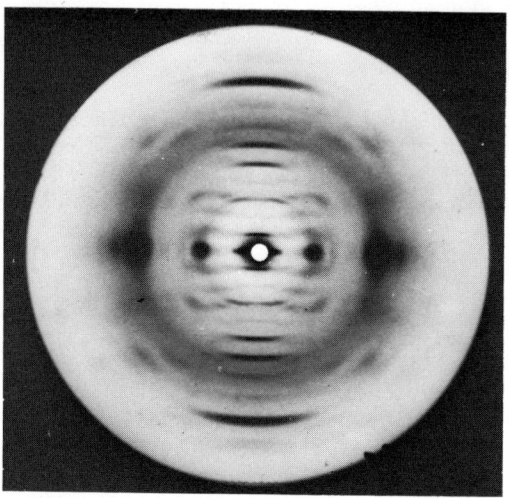

Figure 2. X-ray fiber diagram of intact ovipositor of Megarhyssa lunator. *(Reproduced, with permission, from Ref. 26. Copyright 1980, Academic Press.)*

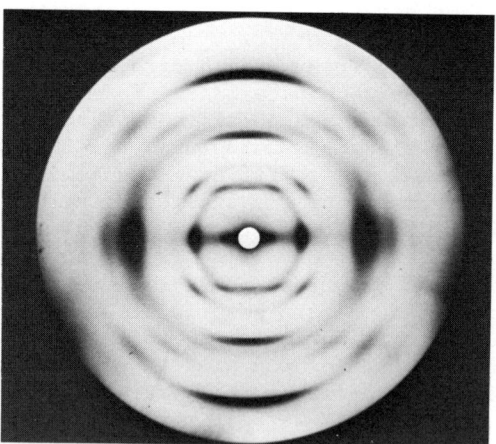

Figure 3. X-ray fiber diagram of deproteinized ovipositor of Megarhyssa lunator. *(Reproduced, with permission, from Ref. 26. Copyright 1980, Academic Press.)*

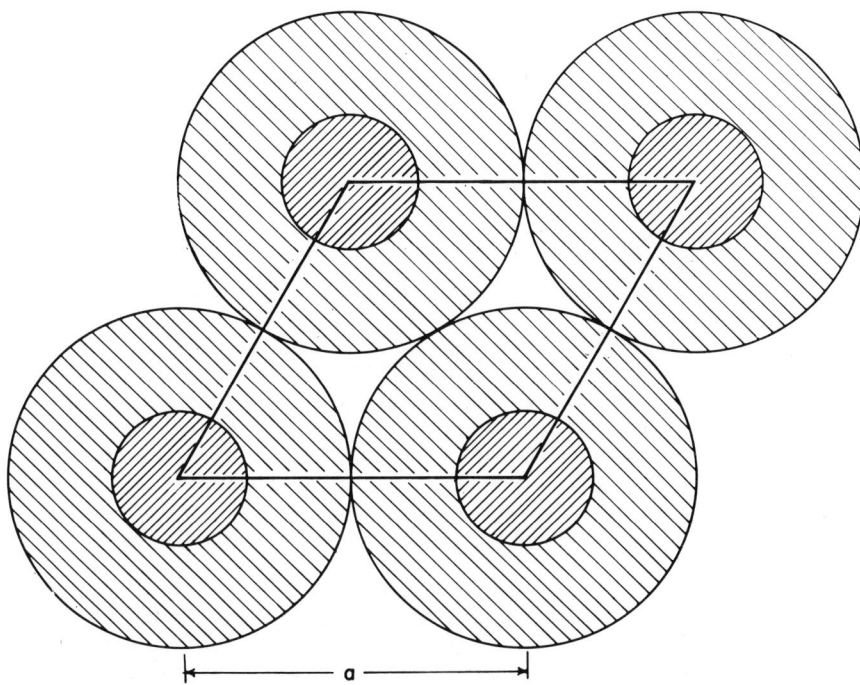

Figure 4. Hexagonal arrangement of core-sheath microfibrils in the proposed structure of Megarhyssa *chitin–protein complex. (Reproduced, with permission, from Ref. 26. Copyright, 1980, Academic Press.)*

The intensity distribution on the protein layer lines suggests a helical arrangement of protein subunits. The strong intensity on layer line 1 is seen on close inspection to be split, i.e. it is off-meridional; and similarly, the intensity at the meridian is split in photographs of tilted specimens. Compared to layer lines 1 and 5, layer lines 2 and 4 are weaker and have their maxima further from the meridian. Layer line 3 is mainly due to the chitin; the protein contribution is weak and the first maxima is even further from the meridian than those on layers 2 and 4. This intensity distribution is suggestive of a helical structure with <u>six</u> units per turn repeating in 3·06nm. Such a structure would have a meridional reflection on the sixth layer line, as is observed, although of course this coincides with the strong meridional on the second layer line of the α-chitin pattern. From the position of the first maximum on the second (protein) layer line the radius of the equivalent point helix is calculated to be 2·42nm, which should correspond approximately to the mid-point of the protein core. This compares well with calculated mid-point radii of 2·51 and 2·76nm for core-sheath models with core diameters of 2·8 and 3·8nm respectively.

From the above arguments we can propose the core-sheath model shown in fig. 5: the chitin core is surrounded by a helix of six protein subunits repeating in 3·06nm. We have shown the subunits as globular proteins, and if the basic subunit is indeed a protein of molecular weight 27,000, then this would have approximately the dimensions shown. However, a more heterogeneous protein population could also be compatible with the proposed symmetry, and the scattering subunits could contain sections of several protein chains, perhaps with some fibrous character. Further work is necessary on the chemistry of the proteins before we can proceed further in this area.

On closer inspection it can be seen that this treatment of the protein data is an over simplification, in that the layer lines identified as 1, 2, 3... above are in fact split into several components. This indicates that the description of the protein structure as a 6_1 (or 6_5) helix repeating in 3·06nm is only an approximation to a structure with a longer repeat. From our measurements of the splitting of the third layer line, a rough estimate of the actual repeat is 15·3nm, or more probably an integral multiple thereof. This longer repeat could arise from a small additional twist to the protein helix, giving it a non-integral character, or the formation of a coiled-coil structure. We are currently working on a number of other chitin-protein complexes, all of which have the same basic 6_1 helical structure repeating in approximately 3·06nm, but have different layer line splittings and long repeats for the overall protein structure. This long repeat may be significant in view of the numerous electron microscope observations that chitin (and also cellulose) microfibrils tend to show a ribbon like morphology with a slow twist.

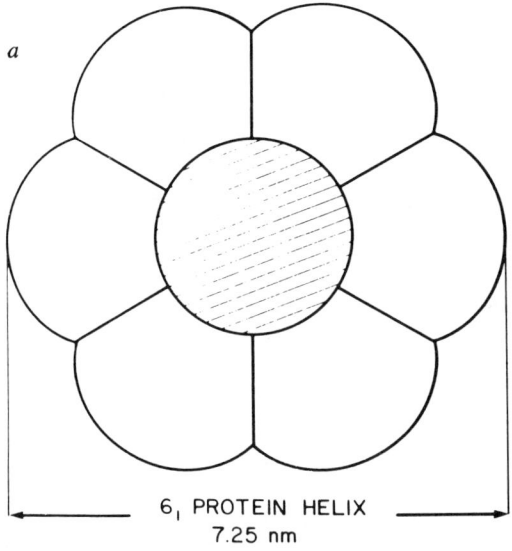
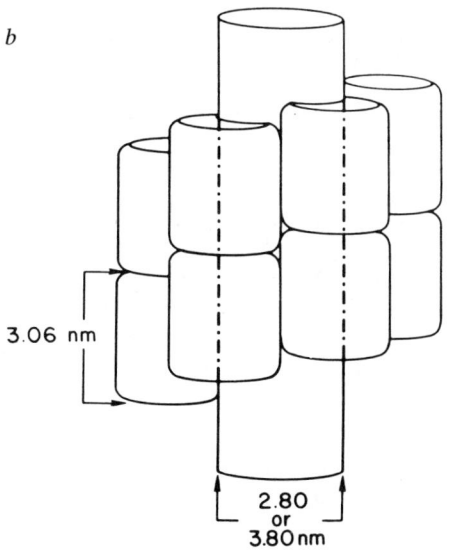

Figure 5. Proposed arrangement of the protein subunits around the chitin core in Megarhyssa *chitin–protein complex. Key: a, axial projection and b, projection perpendicular to fiber axis. (Reproduced, with permission, from Ref. 26. Copyright 1980, Academic Press.)*

Reconstruction of Electron Micrographs

Figure 6 shows an electron micrograph of a transverse section of *Megarhyssa* ovipositor, in which the protein is stained with uranyl acetate and lead citrate. Unstained chitin microfibrils are seen in a matrix of stained protein; the fibrils are arranged hexagonally in the well ordered regions, and have a center to center distance of 7·5 ± 0·5nm. When we examine the protein matrix in more detail it is seen to have a granular appearance, consisting of stained dots which are 2 - 3nm across. In certain areas there is a suggestion of sixfold symmetry to these dots, although this is not nearly as perfect as the fibril arrangement. We have therefore used Fourier reconstruction methods to see if the micrographs contain general evidence of protein substructure, and our preliminary results are shown below.

Figure 7 shows the Fourier reconstruction of a region selected from the micrograph in fig. 6, in which the chitin fibrils form a well ordered hexagonal array. The procedure used calculates the Fourier transform which shows the maxima predicted for the hexagonal lattice, plus additional information or "noise" between these maxima due to the irregularities in the structure. If it can be assumed that these irregularities arise due to artifacts in specimen preparation, then the noise can be ignored or filtered out, and the image can be resynthesized from the hexagonal lattice (hk0) maxima. The filtered image is shown in fig. 7b, based on 46 observed maxima out to $d \simeq 20\text{Å}$, along with the original unfiltered region of the micrograph. The reconstruction shows an array of black dots around the unstained chitin fibril, and this array appears to have approximately sixfold symmetry. Once sixfold symmetry can be assumed it is possible to identify fibrils for which the immediate protein environment is particularly well ordered. Figure 8a shows fifteen such fibrils selected from different regions of the same micrograph. These were then reoriented to optimize the match between them, and then averaged and filtered using the techniques of Frank *et al*. (28) This resultant average is shown in fig. 8b, and the sixfold arrangement of the protein subunits can be seen clearly.

Although caution is necessary to avoid over interpreting such analyses, these results can be taken as conformation of the sixfold arrangement of the protein subunits. It should be noted that the models from the two techniques are not exactly the same; each fibril in figs. 3 and 4 is surrounded by its own group of subunits, whereas the subunits are shared between the fibrils in fig. 7. Further work is necessary to resolve the details of the protein structure, although these two models are not necessarily inconsistent with each other, given the difference in the experimental techniques.

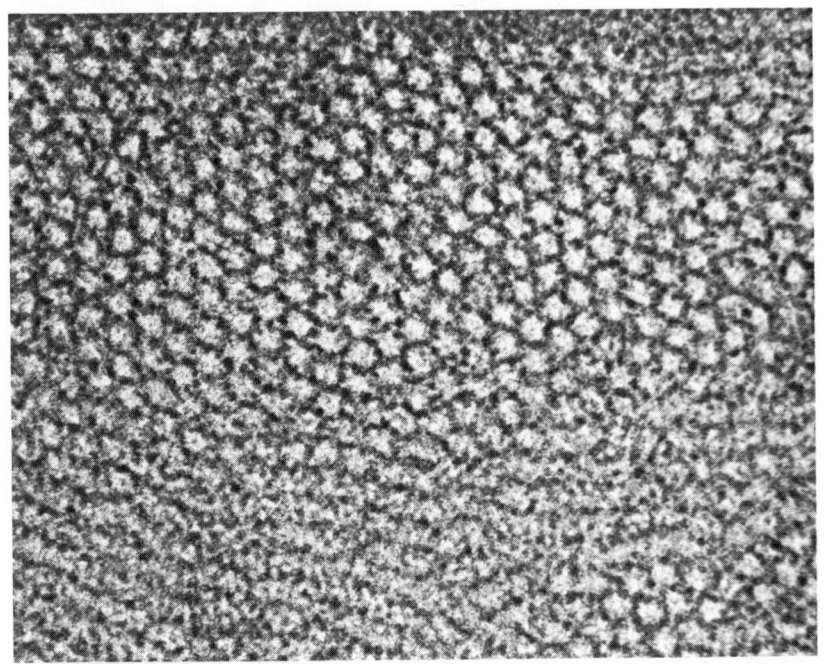

Figure 6. Electron micrograph of transverse section of Megarhyssa *ovipositor stained with uranyl acetate and lead citrate.*

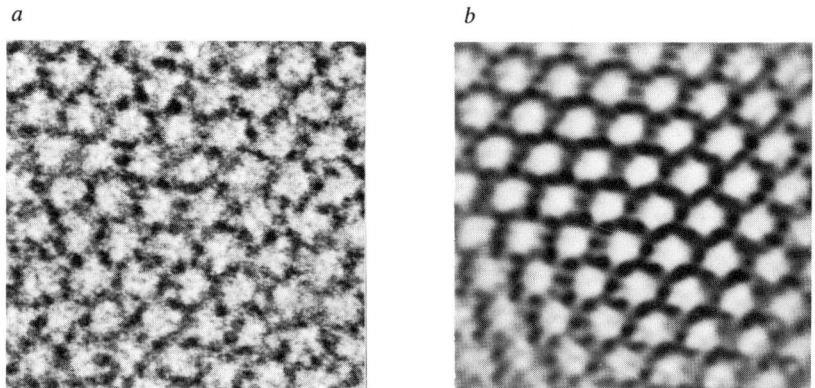

Figure 7. Fourier reconstruction of electron microscope image for Megarhyssa ovipositor. *Key: a, selected well-ordered region and b, reconstructed image.*

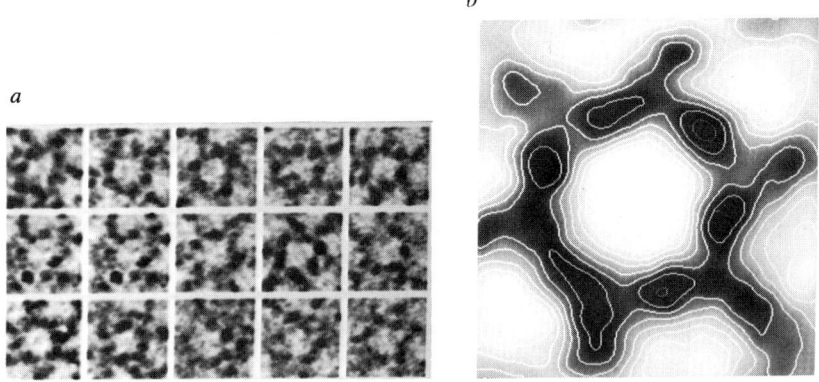

Figure 8. (a) Fifteen selected chitin fibrils with well-ordered surrounding protein and (b) filtered average synthesized from (a).

Acknowledgments

We thank Dr. S. Teraguchi, Dr. A.G. Nielson, and Dr. K.M. Rudall for generously supplying the insect specimens used for this work. We are indebted to Dr. J. Frank of New York State Department of Health, Albany, N.Y. for his collaboration in computer reconstruction of the electron micrographs. Our research was supported by N.S.F. through grants DMR76-82768 and DRM81-07130.

Literature Cited

1. Richards, A.G. "The Integument of the Arthropods"; University of Minnesota Press, Minneapolis, 1951.
2. Rudall, K.M. *Symp. Soc. Exp. Biol.* 1955, 9, 49-71.
3. Neville, A.C. "Biology of the Arthropod Cuticle, Zoophysiology and Ecology Series"; Farner, D.S., Ed.; Springer-Verlag, Berlin, 1975; vol. 415.
4. Giles, C.H.; Hassan, A.; Laidlow, M.; Subramanian, R.V.R. *J. Soc. Dyers Colour* 1958, 74, 647-654.
5. Muzzarelli, R.A.A. "Natural Chelating Polymers"; Pergamon Press, New York, 1973.
6. Hackman, R.H. *Australian J. Biol. Sci.* 1960, 13, 568-577; Hackman, R.H.; Goldberg, M. *J. Insect Physiol.* 1971, 17, 335-347.
7. Attwood, M.M.; Zola, H. *Comp. Biochem. Physiol.* 1967, 20, 993-998.
8. Lipke, H.; Strout, V. *Israel J. Ent.* 1972, 7, 117-128.
9. Brine, C.J.; Austin, P.R. *Comp. Biochem. Physiol. B* 1981, in press.
10. Falk, M.; Smith, D.G.; McLachlan, J.; McInnes, A.G. *Canad. J. Chem.* 1966, 44, 2269-2281.
11. Rudall, K.M. *Advan. Insect. Physiol.* 1963, 1, 257-313.
12. Carlstrom, D. *J. Biophys. Biochem. Cytol.* 1957, 3, 669-683.
13. Dweltz, D. *Biochim. Biophys. Acta* 1961, 51, 283-294.
14. Blackwell, J. *Biopolymers* 1969, 7, 281-298.
15. Gardner, K.H.; Blackwell, J. *Biopolymers* 1974, 14, 1975-2001.
16. Kolpak, F.J.; Blackwell, J. *Macromolecules* 1976, 9, 273-278.
17. Gardner, K.H.; Blackwell, J. *Biopolymers* 1975, 14, 1581-1595.
18. Minke, R.; Blackwell, J. *J. Mol. Biol.* 1978, 120, 167-181.
19. Neville, A.C.; Parry, D.A.D.; Woodhead-Galloway, J. *J. Cell. Sci.* 1976, 21, 73-82
20. Rudall, K.M. in "Conformation of Biopolymers"; Ramachandran, G.N., Ed.; Academic Press, London, 1967; vol. 2.
21. Rudall, K.M.; Kenchington, W. *Biol. Rev.* 1973, 49, 597-632.
22. Rudall, K.M. "The Insect Integument"; Hepburn, H.R., Ed.; Elsevier Scientific Publishing Co., New York, 1976.
23. Hunt, S. "Polysaccharide-Protein Complexes in Invertebrates"; Academic Press, New York, 1970.

24. Strivastava, R.P. *J. Insect. Physiol.* 1971, <u>17</u>, 189-196.
25. Anderson, S.O.; Chase, A.M.; Willis, J.H. *Insect Biochem.* 1973, <u>3</u>, 171-180.
26. Blackwell, J.; Weih, M.A. *J. Molec. Biol.* 1980, <u>137</u>, 49-60.
27. Germinario, L.T.; Blackwell, J.; Frank, J. 1981, paper in preparation.
28. Frank, J.; Goldfarb, W.; Eisenberg, D.; Baker, T.S. *Ultramicrosc.* 1978, <u>3</u>, 283.
29. Germinario, L.T.; Blackwell, J.; Frank, J. *Proc. 38th Ann. EMSA* 1980, <u>38</u>.
30. Atkins, E.D.T., cited by Rudall (ref. 22).

RECEIVED December 14, 1981.

Interaction of Synthetic Polymers with Cell Membranes and Model Membrane Systems: Pyran Copolymer[1]

LALAT K. MARWAHA and DAVID A. TIRRELL
Carnegie–Mellon University, Department of Chemistry, Pittsburgh, PA 15213

We have investigated the interaction of pyran copolymer with erythrocyte ghosts, with intact erythrocytes and with liposomes of pure dipalmitoylphosphatidylcholine (DPPC). Particular attention was given to the importance of divalent metals (especially Ca^{2+}) in promoting the association of pyran copolymer with cell and lipid surfaces. High sensitivity differential scanning calorimetry shows: i) The perturbation of erythrocyte ghost structure by pyran copolymer is not readily "washed out" by treatment with fresh buffer, ii) This perturbation is not due simply to sequestration of metal ions, iii) Pyran copolymer reverses at least partially the effects of Ca^{2+} added to erythrocyte ghosts, and iv) Pyran copolymer raises the transition temperature of DPPC liposomes in the presence of 5 mM Ca^{2+}, but not in the absence of added Ca^{2+}.

The 1:2 copolymer of divinyl ether and maleic anhydride ("pyran copolymer," I) exhibits a broad range of biological activity, and has been discussed elsewhere in this volume and in earlier reviews (1, 2, 3). Of particular interest is the polymer's antitumor activity; after initial clinical testing and then withdrawal from clinical use due to severe toxicity, the drug is again under evaluation as an agent for the treatment of human cancer.

[1] This is Part IV in a series.

[Structure I is a nominal structure for the 1:2 copolymer of divinyl ether and maleic anhydride, after hydrolysis; in fact, the size of the ether ring is a subject of current investigation (4). Potentiometric titration shows that approximately 3 of the 4 carboxyl groups are ionized at pH 7 (1)].

The mechanism of the antitumor action of pyran copolymer has been studied by a number of groups, and there is now ample experimental evidence supporting the hypothesis that this activity is mediated by macrophages (for a review, see ref. (5)). In addition, Fiel and coworkers (6-9) have studied the interaction of the drug with divalent cations, and have presented evidence for the formation of soluble and insoluble complexes. Recently, these workers have speculated that pyran may function as a macromolecular ionophore, activating macrophages via changes in intracellular ion concentrations (6). Schultz, Papamatheakis and Chirigos (10) have observed that tumor cells aggregate around and adhere firmly to pyran-activated macrophages, in vitro, suggesting structural and functional modification of the macrophage membrane, but the nature of the modification is unknown. Finally, the acute toxicity of the partial Ca^{2+} salt of pyran is significantly less than that of the Na^+ salt, and it has been suggested that divalent metal ions may be important in mediating the interactions of pyran with proteins and with cell surfaces (11).

In this paper, we examine the interactions of pyran copolymer with model biomembranes of two kinds: i) the human red blood cell membrane (or red cell "ghost") and ii) multilamellar suspensions (liposomes) of dipalmitoylphosphatidylcholine (DPPC), a pure synthetic phospholipid. Each of these systems offers advantages in studies of polymer-cell surface interaction: The red cell membrane, while complex, is still the most readily isolated and best understood of the membranes of normal human cells, and its molecular architecture is, in a general way at least, typical of such membranes. The pure phospholipids provide a much simpler biomembrane model, with the prospect of yielding more complete interpretation of experimental observations.

The results described in this paper support the suggestion that divalent metal ions promote the interaction of pyran copolymer with membrane and lipid surfaces.

Experimental

Pyran Copolymer. Pyran copolymer used in calorimetric studies of red cell ghosts was NSC-46015, a gift of Dr. David S. Breslow of Hercules, Inc. For studies of pyran-lipid mixtures, the polymer was prepared by radical copolymerization of maleic anhydride and divinyl ether according to the technique of Breslow (1), using 9/1 acetone/tetrahydrofuran mixture as solvent, and azobisisobutyronitrile as initiator. The inherent viscosity of the sample was 0.189 dl/g (0.5 g/100 ml in 0.05M NaCl, 30°C).

^{14}C-labeled copolymer was prepared in similar fashion, using [2,3-^{14}C]-maleic anhydride (Amersham). Two samples were prepared - one of specific activity 0.66 mCi/mmol and the other of specific activity 0.54 mCi/mmol. The specific activity was varied by changing the ratio of labeled and unlabeled maleic anhydride used in the copolymerization.

Calorimetry. Human blood (Type A+) was obtained from the Central Blood Bank of Pittsburgh, PA., and used before becoming clinically outdated. Red cell (erythrocyte) ghosts were prepared by the method of Dodge (12), and suspended in 310 ideal milliosmolar (imosm) buffer, either phosphate (a mixture of 0.155 M NaH_2PO_4 plus 0.103 M Na_2HPO_4) or cacodylate [0.020 M $Na(CH_3)_2AsO_2$, 0.135 M (HCl + NaCl)], at pH 7.4. Pyran copolymer, ethylenediaminetetraacetic acid (EDTA) and $CaCl_2$ were introduced into the ghost suspensions as required by two washings with a solution of the desired composition, also at pH 7.4. In those experiments where the concentration of pyran copolymer was sufficient to depress the buffer pH, readjustment was made by adding aqueous NaOH. Heat capacity profiles of treated and control ghost suspensions were recorded on a Microcal, Inc., MC-1 microcalorimeter, at a heating rate of 60°C/hr. Control and treated suspensions were run on the same day, within 24 hr of ghost isolation.

Polymer-lipid mixtures were prepared by hydrating the dry lipid (L-α-phosphatidylcholine, dipalmitoyl; Sigma Chemical Co.) in 50 mM Tris (tris(hydroxymethyl)aminomethane) buffer, pH 7.4 containing pyran copolymer at a concentration of 1 mg/ml. The final lipid concentration was also 1 mg/ml. Ca^{2+} was added in the form of $CaCl_2$. Thermal transitions of lipid samples were recorded at a heating rate of 12°C/hr.

Binding of Pyran Copolymer to Intact Erythrocytes. Human erythrocytes were isolated from whole blood (Type A+) by centrifugation, washed 4 times with isotonic saline and then 3 times with isotonic buffer, pH 7.4. Phosphate buffer was used in experiments without added Ca^{2+}; cacodylate buffer in experiments with added Ca^{2+}. The cells were then suspended in buffer at the desired concentration (5×10^8 cells/cm^3, as determined using a hemocytometer). [^{14}C]-Pyran copolymer in buffer was added and the mixture incubated for 10 min at room temperature. The cells were then sedimented by centrifugation and the supernatant removed. After 2 washings with buffer (5 ml each), the cell pellet was dissolved in 1 ml of 10% aqueous Triton X-100 and 10 ml of Budget Solve (Research Products International) scintillation cocktail, and the bound radioactivity determined by liquid scintillation counting, using a Beckman LS7000 Counter. Each sample was run in quadruplicate, and duplicate control experiments without cells were run with each.

Results and Discussion

Figure 1 shows heat capacity profiles of human erythrocyte ghosts incubated with increasing concentrations of pyran copolymer in 310 imosm phosphate buffer. The five endothermic transitions of the control sample (bottom curve) are labeled A, B_1, B_2, C and D, in accord with earlier work by Brandts and coworkers (13-16) and from this laboratory (17-20). The A transition is assigned quite securely as a partial denaturation of spectrin, the major cytoskeletal protein on the erythrocyte membrane (13). B_1 and B_2 appear to involve proteins and lipids (protein Bands IV.1 and IV.2 in B_1 and Band III in B_2) (14). The C transition exhibits many of the characteristics of a phospholipid phase transition, and D has been assigned to a protein denaturation process (13). Thus the membrane is resolved into a number of "domains," and one can probe the interactions of a polymeric solute with each of these structures.

Pyran copolymer at a concentration of 3.7 mg/ml causes a marked change in the appearance of the heat capacity profile. In fact, each transition is affected, with the possible exception of D. Although there are small changes in the temperatures of the A, B_2 and C transitions, we feel at this point that the most significant perturbation is that of the B_1 transition, which appears depressed in temperature and in amplitude. Neither poly(acrylic acid) nor poly(methacrylic acid) affects the B_1 transition in this way. Recently, we have observed a similar "flattening" of the B_1 transition by poly(α-ethylacrylic acid) at a repeating unit concentration of 10 mM (21), and we are now pursuing possible further parallels in the biological properties of poly(α-ethylacrylic acid) and pyran copolymer. As the concentration of pyran copolymer is reduced (middle curves in Figure 1), the shift in each transition decreases, and no perturbation is apparent at 20 μg/ml.

In trying to characterize this interaction more fully, we determined the reversibility of the pyran-induced perturbation of the heat capacity profile. Figure 2 shows profiles for control ghosts (the lower curve) and for ghosts incubated with 3.7 mg/ml of pyran and then washed with fresh buffer 3 times - a treatment which reverses completely any effects of pH, ionic strength or divalent ions (14). The upper curve shows that the effect of pyran copolymer is partially, but not completely, reversible under these conditions, in that the B_2 and C transitions remain shifted. A small loss in amplitude of B_1 may also be significant. This is consistent with our quantitative binding experiments (discussed below) in which we have observed that a small but measurable amount of pyran remains associated with whole cells after a comparable treatment with fresh buffer.

Effects of Divalent Ions. Because of the known affinity of pyran copolymer for divalent ions (6-9), it seemed plausible that

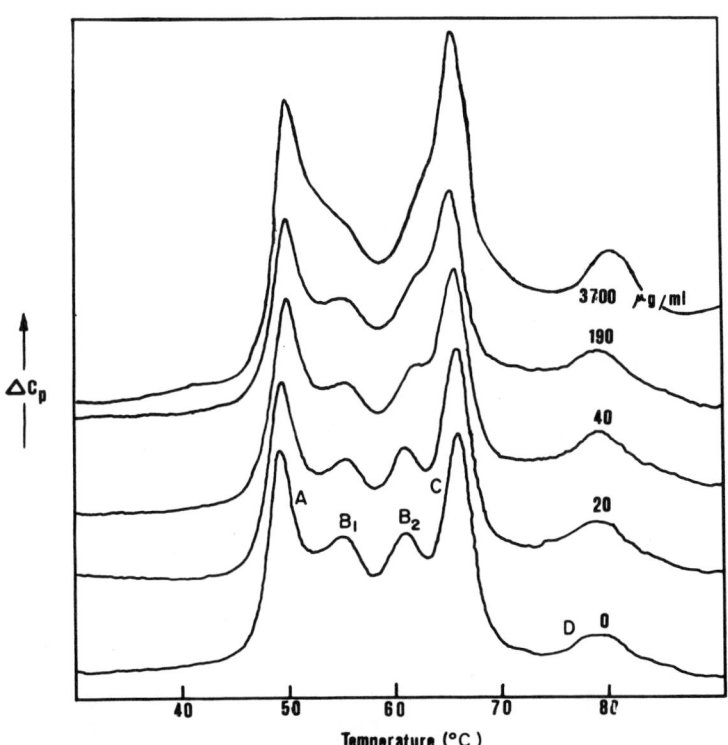

Figure 1. Heat capacity profiles of erythrocyte ghosts suspended in 310 imosm sodium phosphate at pH 7.4. Concentration of pyran copolymer in each suspension is indicated.

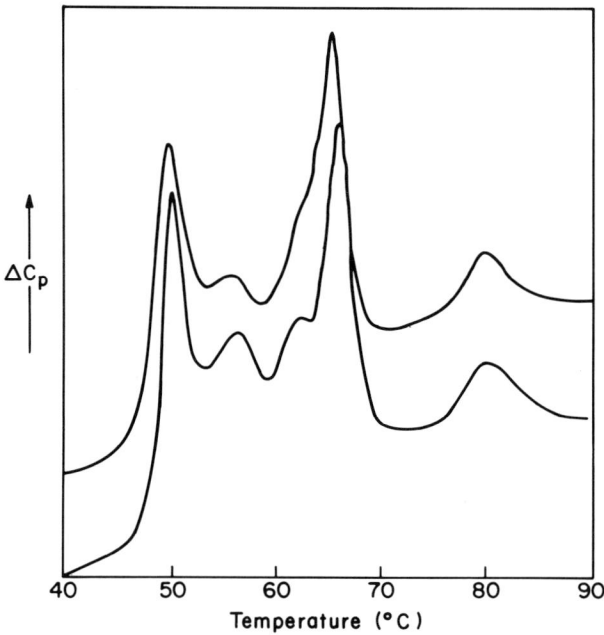

Figure 2. Reversibility of pyran-induced perturbation of heat capacity profile. Key: lower curve, control profile with ghosts suspended in 310 imosm sodium phosphate; and upper curve, ghosts incubated with 3.7 mg/mL pyran copolymer, then washed three times with fresh 310 imosm phosphate buffer.

the observed perturbation of membrane structure might result from a depletion of Ca^{2+} or Mg^{2+} bound to the membrane or introduced in the ghost preparation. A typical preparation contains approximately 0.2 mM Ca^{2+} plus Mg^{2+} by atomic absorption spectroscopy. Figure 3 compares the effects of pyran and EDTA, which should also compete for metal ions. The effects of these two compounds are clearly different, especially in the region of the C transition, which is raised in temperature (from 66° to 68° C) by EDTA, and depressed in temperature by a comparable amount of pyran copolymer. The effects on the B_1 transition also differ markedly. On this basis, it seems clear that the interaction of pyran with the erythrocyte ghost is not simply a sequestration of membrane-bound metal ions, with the formation of soluble polymeric metal salts. On the other hand, one cannot rule out complexation plus association of the complex with the membrane surface.

The importance of divalent ions shows up most clearly in quantitative binding experiments using ^{14}C-labeled pyran copolymer. In these experiments, whole erythrocytes - not isolated membranes - were incubated at room temperature for 10 minutes with the desired concentration of pyran copolymer in buffer. The cells were then sedimented by centrifugation, washed twice with fresh buffer, disrupted with detergent, and the radioactivity determined by liquid scintillation counting. The same procedure was performed as a control in the absence of cells, in order to check for sedimentation of polymer. The results are shown in Figure 4.

Figure 4 is a plot of sedimented radioactivity vs the concentration of pyran in the incubation mixture, in the presence of 5 mM Ca^{2+} and in the absence of added Ca^{2+}. These measurements were done in quadruplicate, with 5×10^8 cells/cm^3, and the error bars are the 95% confidence limits on each value. The effect of Ca^{2+} is striking, in that approximately 100 times as much radioactivity is sedimented in the presence of 5 mM Ca^{2+} as in the absence of added divalent ions. 5 mM represents a reasonable concentration of divalent ions in human plasma, and 60 μg/ml of pyran (the highest concentration shown) would be reasonably obtained in clinical use. It is significant that the increased binding in the presence of Ca^{2+} is observed only when Ca^{2+} is present also in the wash buffer - washing with Ca^{2+}-free buffer reduces the sedimented radioactivity to levels obtained in incubations without added Ca^{2+}.

Complexation with Ca^{2+} seems to cause pyran copolymer to become "sticky," and to adhere to the red cell surface, but the meaning of this result is obscured somewhat by the fact that, at higher concentrations - higher than 60 μg/ml - significant radioactivity is sedimented in the absence of cells, suggesting flocculation of pyran/Ca^{2+} microspheres as observed by Fiel and coworkers. It is possible that we are seeing only a heteroflocculation of microspheres plus red cells, and that the association of pyran with the membrane is nonspecific.

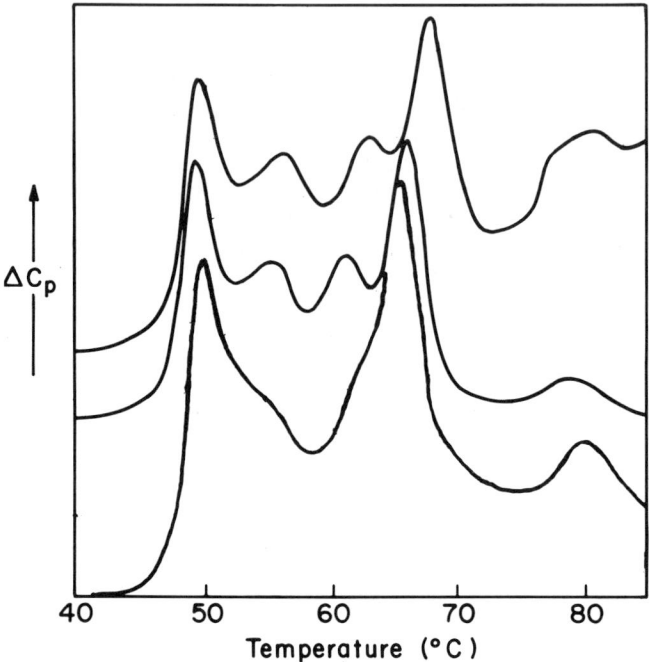

Figure 3. Comparison of effects of pyran copolymer and EDTA in 310 imosm sodium phosphate. Key: center curve, control profile; lower curve, ghosts incubated with 3.7 mg/mL pyran copolymer; and upper curve, ghosts incubated with 3.7 mg/mL EDTA.

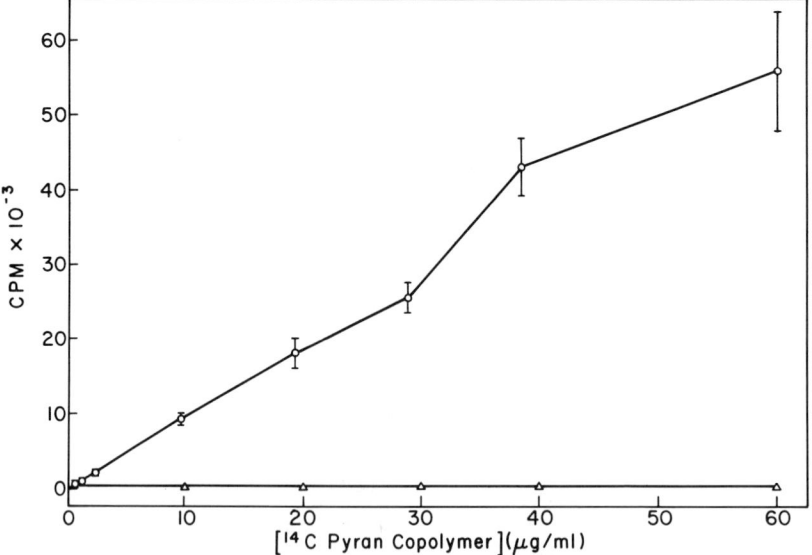

Figure 4. Effect of Ca^{2+} on "binding" of pyran copolymer to intact erythrocytes. Key: △, no Ca^{2+}; and ○, 5mM Ca^{2+}.

Still, the question of structural perturbation of the membrane by pyran copolymer in the presence of physiological concentrations of divalent ions is an important one, so we applied calorimetry to this problem. Figure 5 shows as the top curve the heat capacity profile for ghosts in the presence of 5 mM Ca^{2+}. The A, C and D transitons are apparent, with A centered at about 49.5°C, C at 66°C and D at about 81°C. From Lysko's experiments at lower Ca^{2+} concentrations, it seems that B_1 and B_2 have merged with A and C, respectively (14). As pyran is added, the effects on B_1 and B_2 are reversed, although B_1 is flattened as it is by pyran in the absence of Ca^{2+}. In addition, the C transition is raised in temperature by about 3°. This again is not simply a depletion of membrane Ca^{2+}, since even at 3.7 mg/ml of pyran (more than a 10-fold excess of pyran carboxylate as compared to Ca^{2+}) the profile has not returned to its form in the absence of Ca^{2+}, shown as the bottom curve. Some association is indicated, and this causes significant modification of the thermal transition behavior of the membrane.

The importance of Ca^{2+} is also shown in experiments with pure phospholipids. In this work, we use "liposomes," or multilayer suspensions of a single lipid - dipalmitoylphosphatidylcholine (DPPC) - to approximate the lipid bilayer of the natural membrane. DPPC was chosen because the bulk of the phospholipids in the outer monolayer of the red cell membrane are choline lipids. This then is a much simpler system than is the intact erythrocyte ghost, and it allows us to examine the interaction of the polymer or polymer/metal complex with a single membrane component.

Figure 6 shows our results for DPPC suspended in buffer at a concentration of 1 mg/ml. The control suspension (lower curve) shows the normal pre-transition at 35°C and the main melting transition at 41°C. The main transition, with an enthalpy of about 8 kcal/mole, represents the introduction of a number of gauche bond rotations into the saturated acyl chains of the lipid, which are all trans in the low temperature state. The smaller pre-transition may be associated with reorientation of the phosphocholine headgroup, although there is not complete agreement on this assignment.

Addition of pyran copolymer (at a concentration of 1 mg/ml) to the hydration buffer causes little, if any, perturbation of the transition behavior - perhaps a slight increase in transition half-width - whereas the addition of the same amount of pyran in the presence of 5 mM Ca^{2+} causes each transition to shift up in temperature - the main transition by 1.2°C, and the pre-transition by 4°C. The main transition is also broadened, with the half-width increased from 0.37°C to 0.62°C. Ca^{2+} alone cannot reproduce this effect; in the presence of Ca^{2+} alone each transition is shifted by not more than 0.5°C. We feel that this result suggests a Ca^{2+}-promoted association of pyran copolymer with the lipid surface. The greater perturbation of the "head-group"

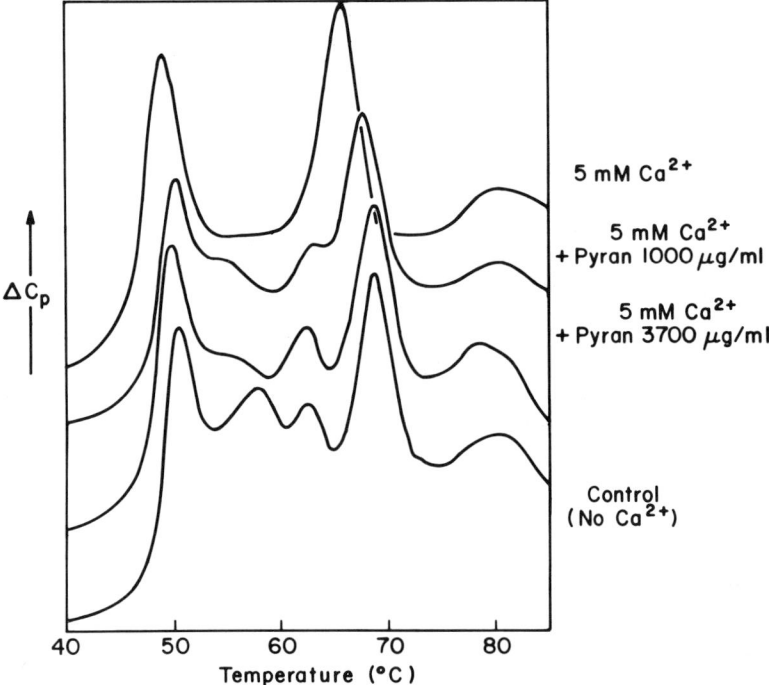

Figure 5. Heat capacity profiles of erythrocyte ghosts suspended in isotonic cacodylate buffer at pH 7.4. The Ca^{2+} and pyran copolymer were added as shown.

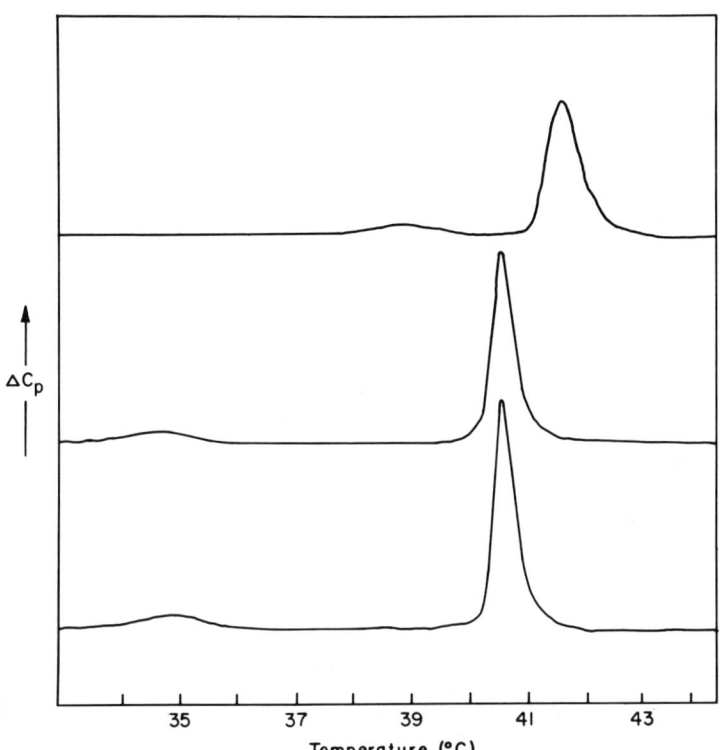

Figure 6. Thermal transition behavior of DPPC in multilamellar suspension in 50mM Tris at pH 7.4. Key: lower curve, DPPC 1 mg/mL; middle curve, DPPC 1 mg/mL + pyran copolymer 1 mg/mL; and upper curve, DPPC 1 mg/mL + pyran copolymer 1 mg/mL + $CaCl_2$ 5mM.

transition vs the hydrocarbon core transition suggests localization of the polymer complex in the head-group region at the surface, and we are pursuing complementary experiments using lipid monolayers in order to test this hypothesis.

Conclusion

Ca^{2+} at concentrations typical for divalent ions in human plasma, promotes the association of pyran copolymer with a pure phospholipid (dipalmitoylphosphatidylcholine) and with intact erythrocytes. This may suggest a more general mediation by divalent ions of the cell surface interactions and pharmacological properties of pyran copolymer.

Acknowledgments

The authors would like to thank Professors William S. Kelley and James F. Williams for use of liquid scintillation counters. This work was supported by the Health Research and Services Foundation, Pittsburgh, PA., and by the Samuel and Emma Winters Foundation. The calorimeter was purchased with the aid of National Science Foundation Grant No. CHE-79-11206.

Literature Cited

1. Breslow, D. S. Pure Appl. Chem. 1976, 46, 103.
2. Ottenbrite, R. M.; Regelson, W.; Kaplan, A.; Carchman, R.; Morahan, P.; Munson, A. in Donaruma, L. G.; Vogl, O., Eds.; "Polymeric Drugs"; Academic Press: New York, 1978, p. 263.
3. Donaruma, L. G.; Ottenbrite, R. M.; Vogl, O., Eds. "Anionic Polymeric Drugs"; John Wiley and Sons: New York, 1980.
4. Freeman, W. J.; Breslow, D. S. Prepr. Am. Chem. Soc. Div. Org. Coat. Plast. Chem. 1981, 44, 108.
5. Kaplan, A. M. in ref. 3, p. 227.
6. Fiel, R. J.; Mark, E. H.; Levine, H. I. in ref. 3, p. 143.
7. Fiel, R.; Mark, E.; Levine, H. J. Colloid Interface Sci. 1976, 55, 133.
8. Levine, H.; Mark, E.; Fiel, R. Arch. Biochem. Biophys. 1977, 184, 156.
9. Levine, H.; Mark, E.; Fiel, R. J. Colloid Interface Sci. 1978, 63, 242.
10. Schultz, R. M.; Papamatheakis, J. D.; Chirigos, M. A. Cell. Immunol. 1977, 29, 403.
11. Morahan, P. S. Am. Chem. Soc. Div. Polymer Chem. Workshop on Water-Soluble and Biomedical Polymers, Bermuda, November, 1979.
12. Dodge, J. T.; Mitchell, C.; and Hanahan, D. J. Arch. Biochem. Biophys. 1963, 100, 119.
13. Brandts, J. F.; Erickson, L.; Lysko, K.; Schwartz, A.T.; Taverna, R. D. Biochemistry, 1977, 16, 3450.

14. Lysko, K. A. E. Ph.D. Thesis, University of Massachusetts, Amherst, MA. USA, 1980.
15. Brandts, J. F.; Taverna, R. D.; Sadasivan, E.; Lysko, K. A. Biochim. Biophys. Acta 1978, 512, 566.
16. Snow, J. W.; Brandts, J. F.; Low, P. S. Biochim Biophys. Acta 1978, 512, 579.
17. Boyd, P. M.; Tirrell, D. A. Prepr. Am. Chem. Soc. Div. Polymer Chem. 1980, 21(1), 188.
18. Tirrell, D. A.; Boyd, P. M. Makromol. Chem., Rapid Commun. 1981, 2, 193.
19. Marwaha, L. K.; Boyd, P. M.; Tirrell, D.A. Prepr. Am. Chem. Soc. Div. Org. Coat. Plast. Chem. 1981, 44, 211.
20. Tirrell, D. A. Prepr. Am. Chem. Soc. Div. Polymer Chem., in press.
21. Marwaha, L. K.; Tirrell, D. A., to be published.

RECEIVED July 14, 1981.

Thermodynamic Characterization of Proflavine Binding to DNA

YOSHIHIRO BABA[1] and CHARLES L. BEATTY
University of Florida, Materials Science and Engineering, Gainesville, FL 32611

AKIHIRO KAGEMOTO
Osaka Institute of Technology, Asahi-Ku, Osaka 535, Japan

The interaction between DNA and proflavine was studied by means of the heating of mixing. From the results, the thermodynamic quantities were estimated for the DNA-proflavine system by an intercalation process. The thermodynamic results show that the intercalation process of DNA-proflavine system would be expected to be dominated by the enthalpy rather than the entropy. In addition, the heat of interaction depends on GC base composition of DNA; the heat of interaction between GC base pairs and proflavine is greater than that between AT base pairs and proflavine.

It is well known that aminoacridine dyes are bound to DNA in two different ways depending on the concentration of dye: the first type corresponds to strong binding[1] and/or intercalation process[2] in which dye molecules are bound to the DNA base at dilute concentration of dye. The other type corresponds to the weak binding[1] and/or stacking process[2] in which dye molecules are bound to the outside of DNA molecules without base specificity as the concentration of dye increases.

---

[1] Permanent address: Osaka Institute of Technology, Asahi-Ku, Osaka 535, Japan.

The interaction for the intercalation process seems to be based on an interaction of flat aromatic dye molecules between the base pairs of DNA as reported by many investigators[1-7]. However, the exact steric location of the intercalated dye and the heat of interaction between DNA and dye have not been established yet.

In this paper, we attempt to elucidate the thermodynamic quantities of the interaction in the dye binding by intercalation process and also explore the effect of GC base composition of DNA on dye binding by measuring the heat of mixing of proflavine (PF) and Cl. perfringens DNA (DNA I), E. coli DNA (DNA II) and M. lysodeikticus DNA (DNA III), respectively because the interaction for intercalation process is based on the interaction between dye and base pairs of DNA.

EXPERIMENTAL

Materials: The DNA's employed were Cl. perfringens DNA (31 % GC, Sigma type XII), E. coli DNA (50 % GC, Sigma type VIII) and M. lysodeikticus DNA (72 % GC, Sigma type XI) and used without further purification. The dye used was proflavine (PF) which was purchased from Aldrich Chemical Co., Inc. All measurements were in 0.1 mol/l tris-HCl buffer solution (pH 7.60).

Apparatus and Procedure: The calorimeter which was used for the measurement of the heats of mixing of DNA and PF was a LKB batch type microcalorimeter (LKB-10700). For calorimetric measurement, the DNA sample was dissolved into the buffer solution, and equal volumes (about 1.2 cm^3) of DNA and dye solutions were mixed. In this case, the concentration of DNA was kept at a definite value of 4×10^{-4} mol/l of phosphate units of DNA while for the dye solution various amounts of dye were used. The DNA concentration was determined from molar extinction coefficient at 260 nm[8] E_p=7400, 6500 and 7000 for Cl. perfringens DNA (DNA I), E. coli DNA (DNA II) and M. lysodeikticus DNA (DNA III), respectively.

In order to obtain information about the binding parameter for the intercalation process, the absorption spectra of solutions containing a constant concentration of PF and varing amounts of DNA were measured by using a spectrophotometer (Perkin-Elmer, Model 552).

RESULTS AND DISCUSSION

Spectral measurement

Typical absorption spectra of solutions containing a definite concentration of PF and various amounts of DNA are shown in Figures 1(a), (b) and (c). As seen in Figures 1, the wavelength of maximum absorbance for each system shifts to red as the

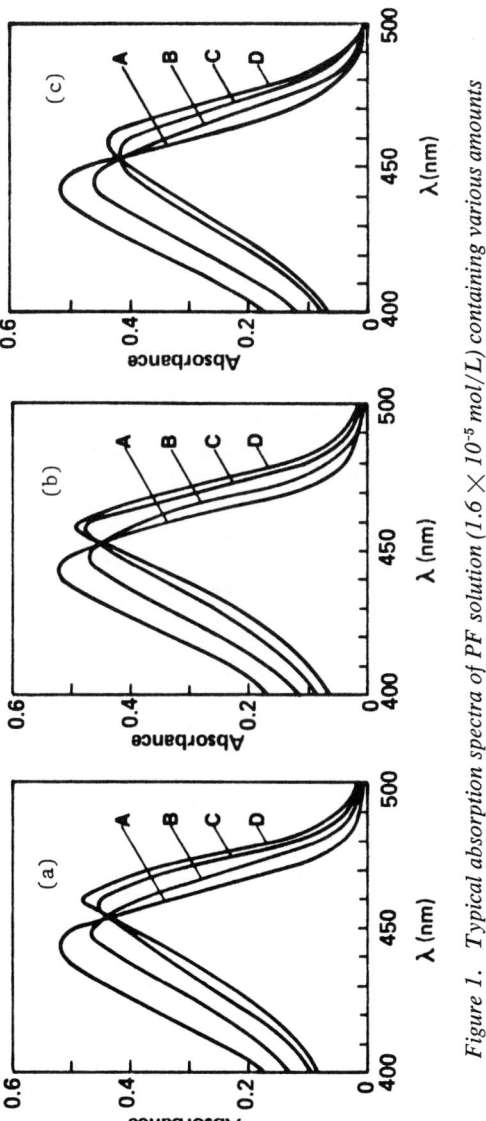

Figure 1. Typical absorption spectra of PF solution (1.6 × 10⁻⁵ mol/L) containing various amounts of DNA.

(a) DNA II–PF system shows DNA concentrations: A, 0; B, 5.36 × 10⁻⁵; C, 1.36 × 10⁻⁴; and D, 5.30 × 10⁻⁴ mol/L. (b) DNA I–PF system shows DNA concentrations: A, 0; B, 3.92 × 10⁻⁵; C, 9.79 × 10⁻⁵; and D, 3.92 × 10⁻⁴ mol/L. (c) DNA III–PF system shows DNA concentrations: A, 0; B, 4.01 × 10⁻⁵; C, 1.15 × 10⁻⁴; and D, 4.95 × 10⁻⁴ mol/L.

concentration of DNA increases and well-defined isosbestic points for each system are located at about 454 nm which is compatible with the previous result[9] and results reported by Lee et al.[10]. The percentage of dye bound to DNA was calculated as described by Peacocke and Skerrett[11]. The plots of r/C_f versus r for each DNA-PF system are shown in Figures 2(a), (b) and (c). In this case, r is the amount of PF bound per nucleic acid phosphate, and C_f is the concentration of free dye. From those plots the binding parameters were estimated and are given in Table I. It is apparent that the binding parameters for each system are in fair agreement, showing that there is no base composition of DNA selectively in binding between DNA and PF.

Heat of Mixing

The heats of mixing of DNA and PF measured at 298.15 K are found to be exothermic, indicating the attractive interaction between DNA and PF molecules. The results are shown in Figure 3, where the heat of mixing per mole of dye, ΔH^M is plotted against the mole ratio of PF to DNA, [dye]/[DNA]. ΔH^M, for each system shows the sigmoidal curve such as ΔH^M increases (takes a smaller negative value) as [dye]/[DNA], that is, PF concentration increases. Those results are compared with the successive stages of the intercalation process as described by Fredricq and Houssier[12]. When [dye]/[DNA] is less than about 0.05, the intercalated dye molecules do not interact with one another, in the range from 0.05 to 0.13 of [dye]/[DNA] the intercalated dye molecules do interact, and then dye molecules begin to bind to the side of DNA molecules when [dye]/[DNA] is more than 0.13.

To obtain the heat of interaction between DNA and PF from calorimetric results, the following analysis is used.

Thermodynamic Quantities

Assuming that the DNA-dye complex is formed by following reaction process between DNA and dye,

$$\text{DNA} + \text{dye} \rightleftharpoons (\text{DNA} - \text{dye}) \tag{1}$$

the binding constant, K, is expressed as

$$K = \frac{r}{C_f (n - r)} \tag{2}$$

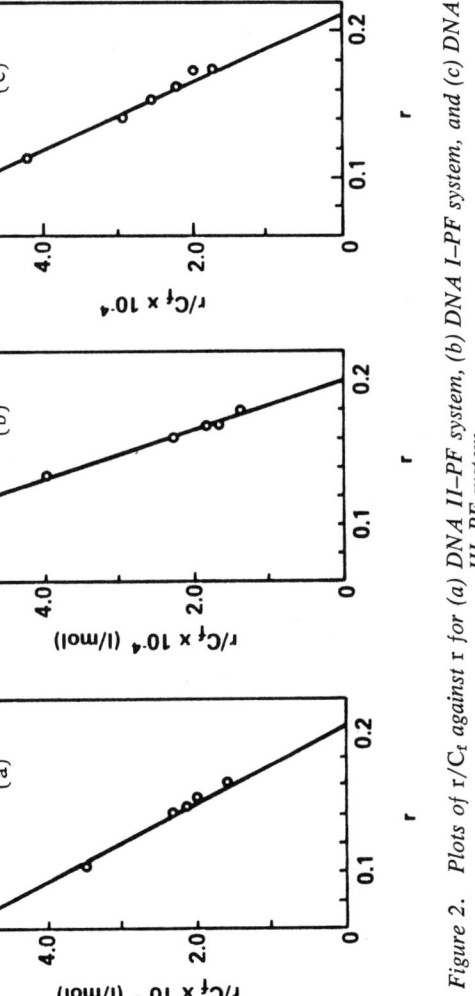

Figure 2. Plots of r/C_f against r for (a) DNA II–PF system, (b) DNA I–PF system, and (c) DNA III–PF system.

Table I The binding parameters for the intercalation process of the DNA-PF complexes estimated from spectrophotometry

DNA	$K \times 10^{-5}$ [a]	n [b]
DNA I (GC 31 %)	5.6	0.20
DNA II (GC 50 %)	4.1	0.20
DNA III (GC 72 %)	6.0	0.21

[a] Binding constant
[b] Number of binding sites per DNA phosphate

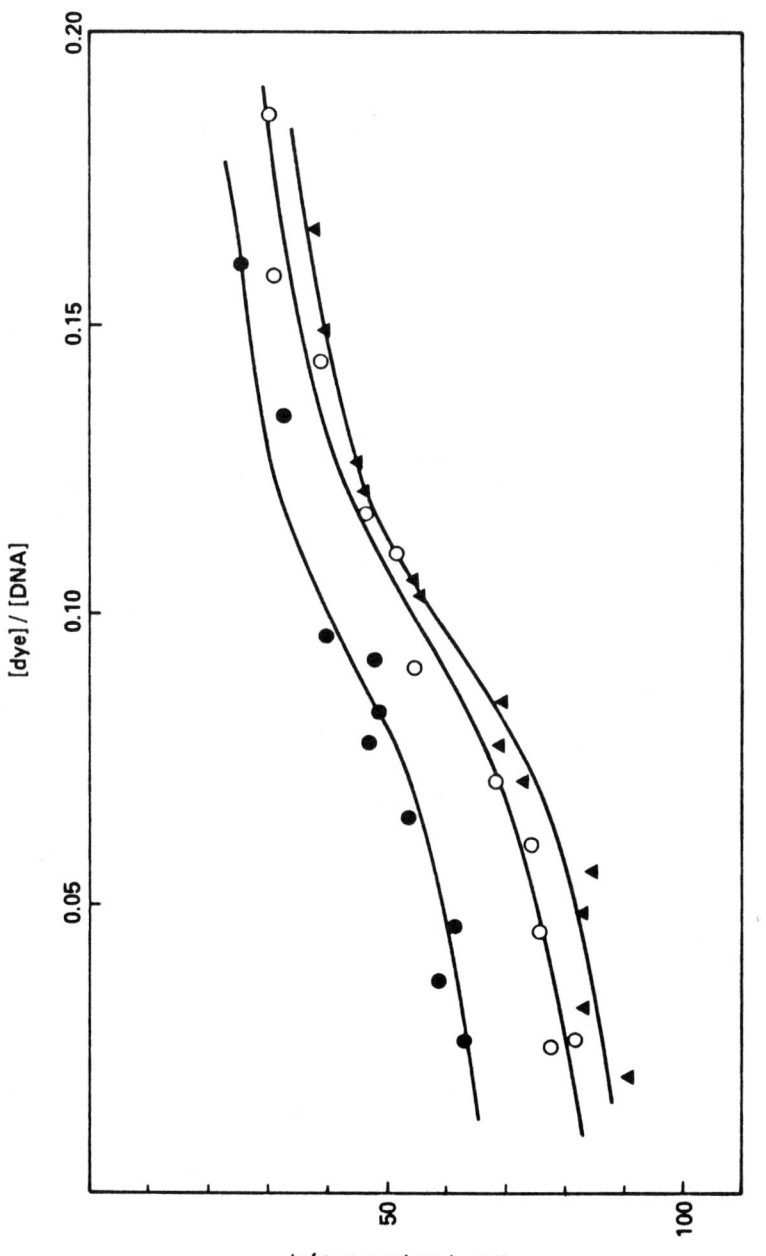

Figure 3. Plots of the heats of mixing, ΔH^M, against the mole ratio of PF to DNA, [dye]/[DNA], for DNA I–PF system (●), DNA II–PF system (○), and DNA III–PF system (▲).

where n is the number of binding sites per DNA phosphate, r the amount of bound dye per DNA phosphate and C_f the concentration of free dye.

From the experimental results of the heat of mixing, the heat of interaction of DNA-dye complex per mole of dye, ΔH, can be expressed as

$$\Delta H = \frac{\Delta Q}{C_b V} \quad (3)$$

Where ΔQ is the observed heat, C_b the concentration of bound dye which is correlated with r and the concentration of DNA phosphate P, such as $C_b = rP$, and V the total volume of the mixture of DNA and dye solutions. The total concentration of dye, C, is the sum of C_b and C_f;

$$C = C_b + C_f \quad (4)$$

From Eqs. (2) to (4), ΔQ will be represented by following complicated function of C, in which the negative sign of the right side must be adopted because correct zero value of $\Delta Q/V$ is given at zero of the concentration of dye, C.

$$\frac{\Delta Q}{V} = \frac{\Delta H}{2} [Pn + \frac{1}{K} + C - \sqrt{(Pn + \frac{1}{K} + C)^2 - 4PnC}\ ] \quad (5)$$

According to Eq. (5) we are able to calculate the value of $\Delta Q/V$ for a given value of C by using both the values of K and n estimated from spectrophotometry. The ΔH value which gives the best fit between the calculated value and the experimental one of $\Delta Q/V$ is adopted. The ΔH values adopted are about -70, -85 and -90 kJ/mol for DNA I, DNA II and DNA III, respectively. But the calculated value of $\Delta Q/V$ deviates from the experimental value in the concentration region of dye more than about 3×10^{-5} mol/l as shown in the broken lines in Figure 4. Those deviations may be a result of K and n values estimated from spectrophotometry in which the concentration of DNA is different from that in calorimetry.

To improve those discrepancies, K and n can be treated as variable parameters and adjusted in such a way that the best fit between experimental and calculated values of $\Delta Q/V$ is obtained. These quantities are plotted against C as a solid line in Figure 4 and the values of ΔH, K and n estimated are listed in Table II. The K values for each system estimated from calorimetry are in agreement with those from spectrophotometry, but the n values estimated from calorimetry are a little lower than those from spectrophotometry. The decrease in n values may be ascribed to the intermolecular interaction of DNA which increases with the concentration of DNA and brings about the decrease of effective binding sites of DNA.

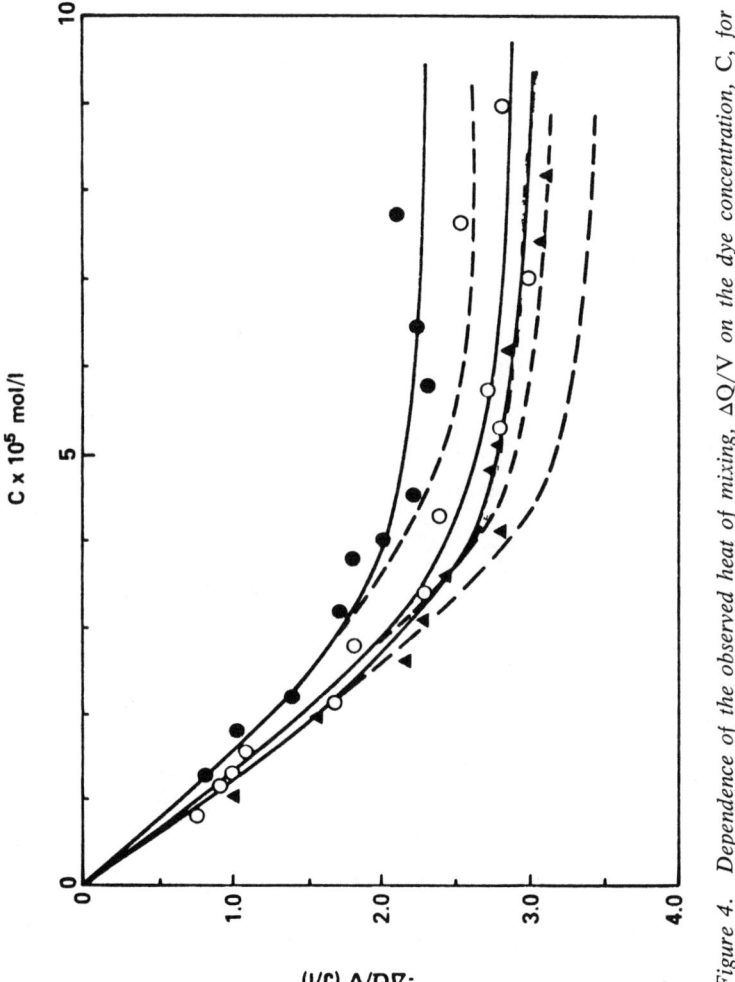

Figure 4. Dependence of the observed heat of mixing, $\Delta Q/V$ on the dye concentration, C, for DNA–PF system. Key: (●) DNA I–PF system; (○) DNA II–PF system; and ▲, DNA III–PF system; theoretical $\Delta Q/V$ curve calculated with Equation 5 using K and n values in Table I (– – –) and in Table II (———).

Table II The thermodynamic quantities for intercalation process of DNA - PF system estimated from calorimetry at 298 K

DNA	$K^{a)}$ / l mol^{-1}	ΔG / kJ mol^{-1}	ΔH / kJ mol^{-1}	ΔS / J K^{-1} mol^{-1}	$n^{b)}$
DNA I (GC 31 %)	5.6 ×10^5	-33	-70	-120	0.17
DNA II (GC 50 %)	4.1 ×10^5	-32	-85	-180	0.18
DNA III (GC 72 %)	6.0 ×10^5	-33	-90	-190	0.18

mol refers to the number of moles of PF
a) Binding constant
b) Number of binding sites per DNA phosphate

The free energy change, ΔG, and the entropy change, ΔS, for each system are easily calculated from the following equations,

$$\Delta G = -RT \ln K, \qquad \Delta S = (\Delta H - \Delta G)/T \qquad (6)$$

and are listed in the third and fifth columns in Table II, respectively.

As seen in Table II, the values of ΔH for each system are different and decrease with decreasing GC base composition of DNA although the values of ΔG are nearly equal. The ΔH dependence on GC base composition of DNA seems to be based on difference of the heat of interaction between GC base pairs and PF and between AT base pairs and PF.

Base composition effect of heat of interaction

Assuming that the three types of DNA/dye interaction (i.e. between GC pairs, AT pairs and GC-AT base pairs) occur in numbers given by the probability of existence of such pairs when DNA molecule corresponds to a random chain, then the heat of interaction, ΔH, is represented as

$$\Delta H = n_A^2 \Delta H_1 + n_B^2 \Delta H_2 + n_A n_B \Delta H_3 \qquad (7)$$

where ΔH_1, ΔH_2 and ΔH_3 are the heats of interaction between GC base pairs and dye, AT base pairs and dye and GC-AT base pairs and dye, respectively, and n_A and n_B are fractions of GC and AT base pairs in DNA molecule. Since n_B is equal to one minus n_A, ΔH is also written as the function of n_A,

$$\Delta H = (\Delta H_1 + \Delta H_2 - 2\Delta H_3)n_A^2 + 2(\Delta H_3 - \Delta H_2)n_A + \Delta H_3 \qquad (8)$$

The plots of ΔH estimated from calorimetry versus GC fraction of DNA are shown in Figure 5, and the solid line in Figure 5 shows the best fit curve of Eq. (8). From this curve, we can obtain $\Delta H_1 = -96$, and $\Delta H_2 = -50$ kJ/mol.

Conclusion

From the thermodynamic quantities estimated from calorimetry, the intercalation process of DNA - PF system would be expected to be dominated by the enthalpy rather than entropy because of negative value of the heat of interaction and positive value of entropy change for the interaction between DNA and PF. The heat of interaction between DNA and PF is greater than that between DNA and acridine orange[13]. This greater heat of interaction may be based not only on interaction between base pairs of DNA and PF but on hydrogen bonding between phosphate group with negative charge of DNA and amino group on 3 and/or 6 position of PF as reported by Peacocke et al.[14].

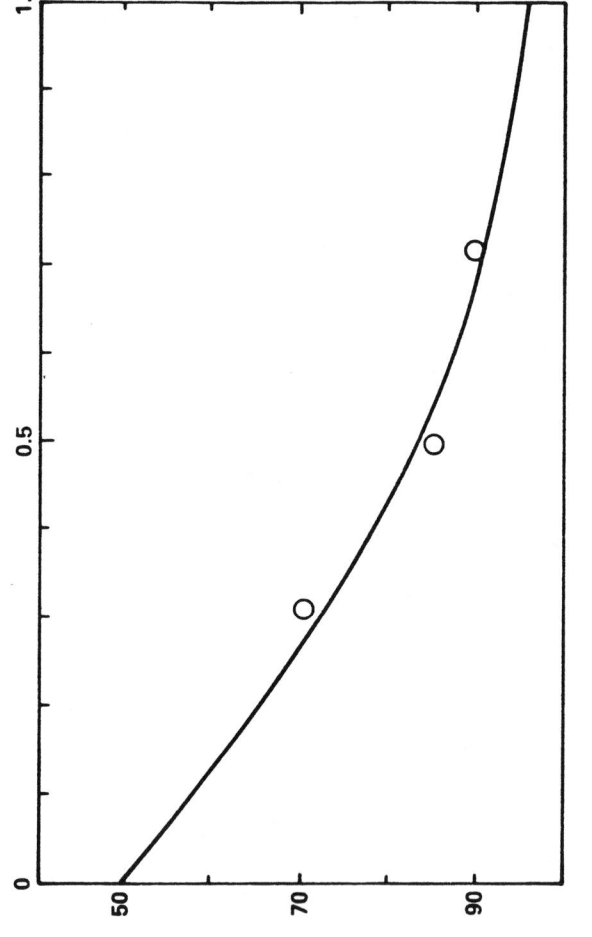

Figure 5. Dependence of the heat of interaction, ΔH, on GC base composition of DNA.

In addition the dependence of the heat of interaction of GC base concentration of DNA may be based on the difference of the heat of interaction between GC base pairs and PF and AT base pairs and PF because the heat of interaction of GC base pairs and PF is greater than that between AT base pairs and PF.

Acknowledgment

The authors (Y. Baba and C. L. Beatty) would like to acknowledge partial support of this work by the Biomedical Center of Excellence, University of Florida.

Literature Cited

1. A. Blake and A.R. Peacocke, Biopolymers, 6, 1225 (1968).
2. L.S. Lerman, J. Mol. Biol., 3, 18 (1961).
3. A. Blake and A.R. Peacocke, Biopolymers, 5, 39 (1967).
4. A Blake and A.R. Peacocke, Biopolymers, 5, 871 (1967).
5. D.M. Neviller, Jr., and D.R. Davies, J. Mol. Biol., 17, 57 (1966).
6. B.J. Gardner and S.F. Mason, Biopolymers, 5, 79 (1967).
7. N.F. Gersch and D.O. Jordan, J. Mol. Biol., 13, 138 (1965).
8. G. Felsenfeld and S.Z. Hirschman, J. Mol. Biol., 13, 407 (1965).
9. S. Tanaka, Y. Baba, A. Kagemoto and R. Fujishiro, Makromol. Chem., 181, 2175 (1980).
10. C.H. Lee, C.T. Change and J.W. Wetmur, Biopolymers, 12, 1099 (1973).
11. A.R. Peacocke and J.N. Skerrett, Trans. Farady Soc., 52, 261 (1956).
12. E. Fredericq and C. Houssier, Biopolymers, 11, 2281 (1972).
13. Y. Baba, C.L. Beatty and A. Kagemoto, in preparation.
14. N.J. Pritchard, A. Blake and A.R. Peacocke, Nature (London), 212, 1360 (1966).

RECEIVED February 5, 1982.

ANTICANCER APPLICATIONS

15

Polymers for Potential Cancer Therapy: A Brief Review

CHARLES G. GEBELEIN

Youngstown State University, Department of Chemistry, Youngstown, OH 44555

A polymeric drug is a polymer that contains a drug unit either as part of the polymer backbone, as a terminal group, or as a pendant unit off the polymer backbone. In some cases, a polymer can function as a drug even though there are no low molecular weight analogs of that polymer. Polymeric drugs can be subdivided into (a) insoluble polymers, (b) soluble polymers and (c) directed polymers. Many examples have been reported of polymers which exhibit antineoplastic activity. In this review, antineoplastic polymers are subdivided into the classes: (a) polyanions, (b) nucleic acid analogs and (c) all other systems. Examples are cited for each of these different catagories of antineoplastic polymers.

The potential use of polymeric materials in treating disease is now well documented by many research papers and several reviews (1-10). Basically, there are three ways that polymers can be used in disease therapy: (a) controlled release systems, (b) mechanical or diffusion controlled pumps and related devices, and (c) polymeric drugs. A number of other papers in this volume will consider the controlled release of chemotherapeutic agents and some recent reviews (11-14) have covered the topics (a) and (b). This paper will concentrate only on the polymeric drug approach to the treatment of neoplastic diseases.

Polymeric Drug Therapy

The basic concept of polymeric drug therapy has been reviewed recently (1,3,5,9). A polymeric drug is a polymer which contains a drug unit either as part of the polymer backbone, as a terminal unit, or as a pendant unit off the polymer backbone. In some cases, the polymer itself functions as the therapeutic agent,

0097-6156/82/0186-0193$5.00/0
© 1982 American Chemical Society

whereas the low molecular weight analogs are inactive. There are three distinctly different types of polymeric drugs: (a) insoluble polymers, (b) soluble polymer systems, and (c) directed polymer systems. In all three types it is assumed that the polymers would be biologically active rather than being a source of slow release of the drug. Generally the insoluble polymeric drugs would be implanted at or near the disease site and would interact with this disease. Since the material would be fairly insoluble in the body fluids, there would be very little interaction with other body tissues or organs. Although this technique would aid in reducing the toxic side effects of the drug such implantation would generally require surgery. Drug activity would be prolonged largely because of the polymer's lack of solubility which would retard migration and excretion. The insoluble polymeric drugs, which are usually the easiest of the three types to prepare, would thus have some advantage over a simple drug although this advantage would probably be of limited value since surgical implantation would usually not be desirable.

The soluble polymeric drugs, on the other hand, would be capable of solution in the body fluids and could be introduced orally or by injection. In simple, non-polymeric drugs, the only way to change the solubility characteristics would be to modify the chemical structure of the drug, itself although this might also change the therapeutic characteristics as well. Generally in a soluble polymeric drug system, the therapeutic and solubilizing parts of the molecule are located in different parts of a copolymer chain and modification of one portion would have little effect on the other. Such copolymers could be prepared by copolymerization of a monomer containing the drug unit with another monomer containing groups which would make the copolymer soluble in the body fluids. Unfortunately, since the resulting polymeric drug is soluble in the body fluids, it could migrate throughout the body about as well as the original drug and could still show toxic side effects. These systems could, however, exhibit prolonged activity and could be active at a lower total concentration since the fact that the drug is polymeric would increase the concentration greatly on a localized scale in the solution. From the standpoint of ease of usage, these would be better than the insoluble polymeric drugs.

The directed polymeric drugs would consist of at least three functional units although lesser number of units might suffice. Basically directed polymeric drug would contain (a) the drug unit, (b) a unit to make the copolymer soluble, and (c) a unit to direct the copolymer to a specific type of diseased tissue or organ. In the case of an antineoplastic polymer, this directing unit would be some unit that would tend to migrate selectively into the tumorous tissue and would thereby carry the copolymer along with it. These systems could be administered orally or by injection of a solution and, since the polymer would go selectively to the diseased site, such directed polymeric drugs would have minimal interactions with other organs which would result in the reduction of the toxic side effects. Directed polymeric drugs would be the most effective type of polymeric drugs but would be the most difficult type to design.

Numerous antineoplastic polymers have been prepared with varying degrees of activity. For convenience, these polymers are divided here into (a) polyanions, (b) nucleic acid analogs, and (c) other systems. Most of these polymers are of the insoluble or soluble class but much recent advances have been made toward developing directed polymeric drugs. Several workers have utilized controlled release systems for the release of anti-tumor drugs although this is beyond the scope of this present review (11,15).

Polyanionic Antineoplastic Polymers

Numerous polyanionic materials have been examined for potential biological activity (6,16). The rationale for this approach probably arises from the fact that many naturally-occurring, biologically-active polymers, such as heparin, DNA, various proteins, alginic acid and pectin, are polyanionic polyelectrolytes. Statolon, which is a polyanionic, polysaccharide, has been shown to supress leukemia in mice (17). The first synthetic polyanionic polymer to be examined extensively for antitumor activity appears to be sodium poly(vinyl-sulfonate) (I) which inhibited the following types of tumors in mice: adenocarcinoma 775, Ehrlich (ascites), Krebs 2 carcinoma (ascites), L1210 lymphoid leukemia, L5178 lymphatic leukemia and sarcoma 180 (6,18). Subsequently it was shown that poly(acrylic acid) (II), poly(methacrylic acid) (III), and the copolymer of ethylene-maleic anhydride (IV) had antineoplastic activity (19). Unfortunately, most of these polymers were extremely toxic and could not be tested clinically in humans.

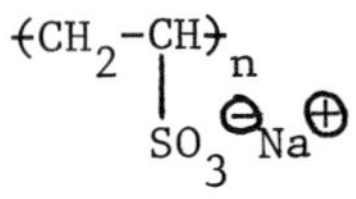

$+CH_2-CH+_n$
 |
 $SO_3^{\ominus} Na^{\oplus}$

(I)

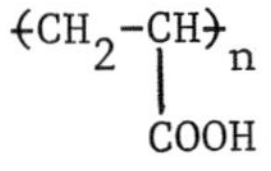

$+CH_2-CH+_n$
 |
 COOH

(II)

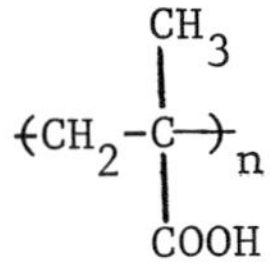

$\quad\quad CH_3$
$\quad\quad |$
$+CH_2-C+_n$
$\quad\quad |$
$\quad\quad COOH$

(III)

$+CH_2CH_2-CH-CH+_n$
$\quad\quad\quad\quad\quad | \quad\; |$
$\quad\quad\quad\quad\; O{=}C \quad C{=}O$
$\quad\quad\quad\quad\quad\;\; \backslash O /$

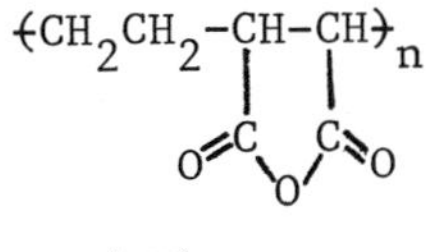

(IV)

One of the most interesting polymers of this type is the 1:2 divinyl ether: maleic anhydride copolymer (DIVEMA) (V) and the hydrolysis product (VI). Polymer (V) was first reported by Butler in 1960 (20) and has since been shown to exhibit an incredibly wide spectrum of biological activity including antitumor, antiviral, antibacterial, antifungal, anticoagulant and antiinflammatory activity (6). While this polymer is shown below as the six-membered pyran ring, this structure has not been fully proven and some other cyclopolymerization products do give five-membered rings instead (16,21).

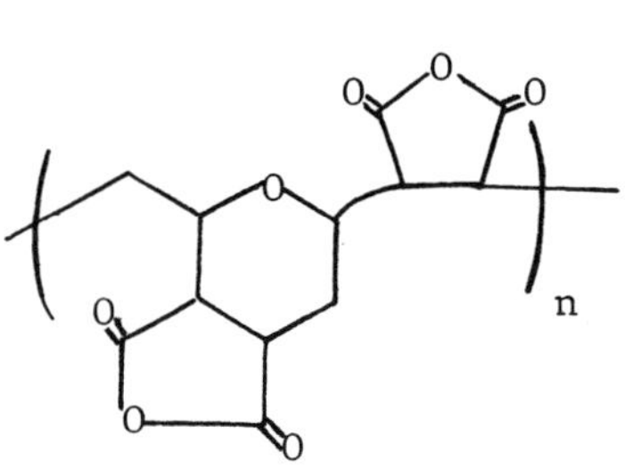

(V) 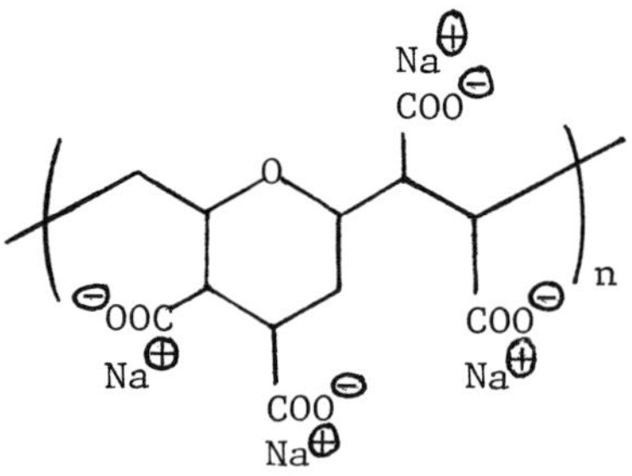   (VI)

Polymer (VI) has been shown to have antitumor activity against adenocarcinoma 755, Dunning ascites leukemia, Friend leukemia virus, and Lewis lung carcinoma. In the latter case, polymer (VI) showed activity about equal to that of cyclophosphamide (an alkylating agent) and was more effective than 6-mercaptopurine (an antimetabolite) (6). The DIVEMA polymer (V) is also active against some cancer causing viruses such as Friend leukemia, Moloney sarcoma and Rauscher leukemia (6).

There has been much speculation concerning the mechanism of tumor inhibition by these polyanionic systems. Many polyanions induce interferon production (22) and this fact has also been suggested as the source of the activity. More recently, it has been observed that DIVEMA is not cytotoxic to tumor cells (23) and that only those polymers that activated macrophages showed antineoplastic activity (24). Inhibition via macrophages is believed to be the mode of action for DIVEMA (V) if not for all polyanionic polymers (16).

Nucleic Acid Analogs as Antineoplastic Polymers

A wide variety of polymers have been prepared containing a nucleic base unit on the polymer chain. In most cases this was achieved by polymerizing a monomeric derivative of the nucleic base. The resulting polymers are similar in some respects to DNA or RNA and could have utility in cancer treatment by interfering with tumor cell replication.

15. GEBELEIN   *Polymers for Potential Cancer Therapy*

The most extensive work in this area has been on the vinyl-type derivatives of uracil and adenine and this research has been reviewed recently (8,25,26). These monomers include 1-vinyl-uracil (VII) and 9-vinyladenine (VIII) (27,28); methacryloylethyl derivatives of uracil (IX) and adenine (X) (29,30); 1-(N-vinylcarbamoyl)uracil and its 5-fluoro- derivative (XI) (31-34); 9-(N-vinylcarbamoyl)-methylthiopurine (XII) (34-36); 9-vinyl-6-alkylthiopurine (XIII) (37); and styrene derivatives of uracil (XIV) and adenine (XI) (38,39).

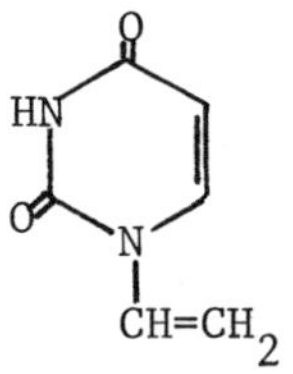

(VII)

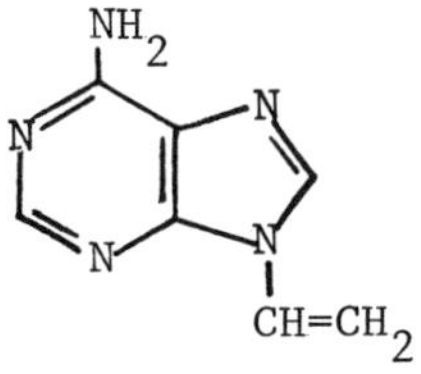

(VIII)

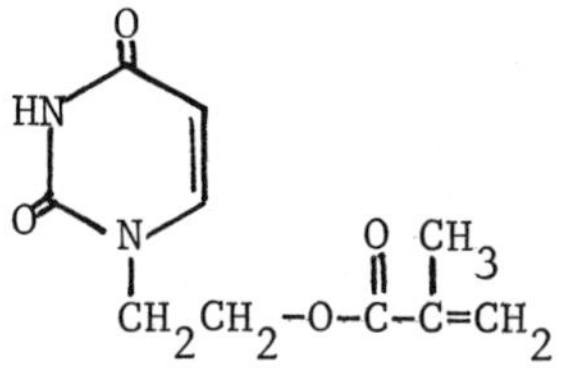

(IX)

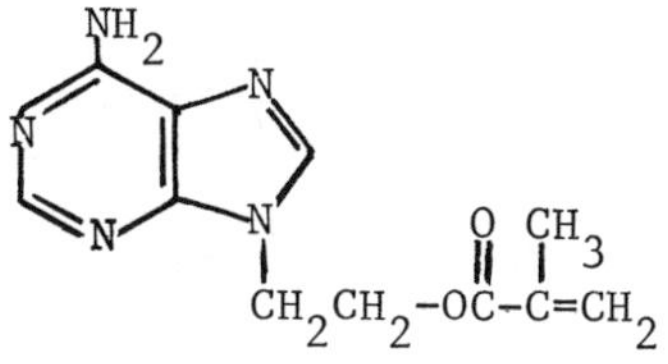

(X)

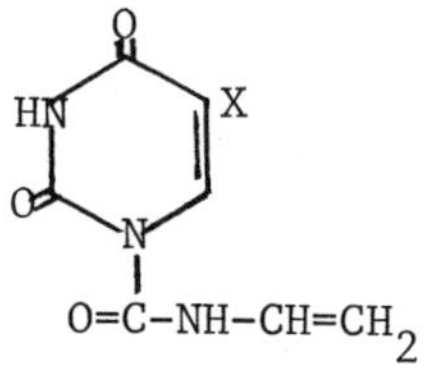

(XI)

(a: X = H)
(b: X = F)

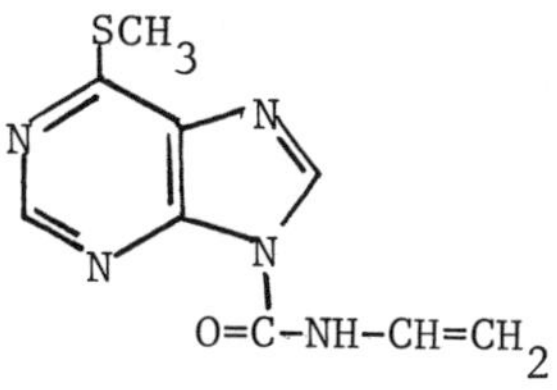

(XII)

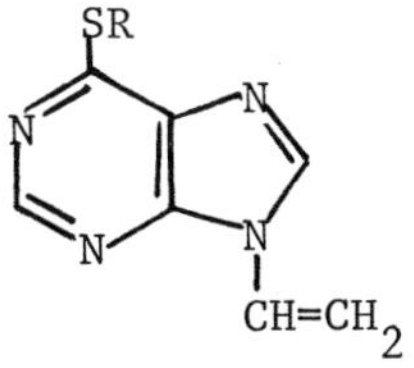

(XIII)

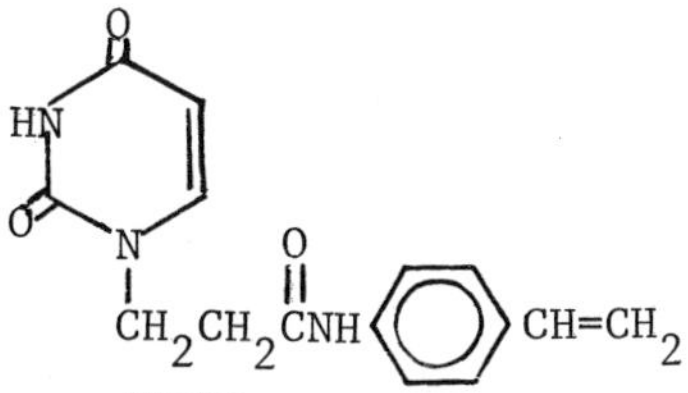

(XIV)

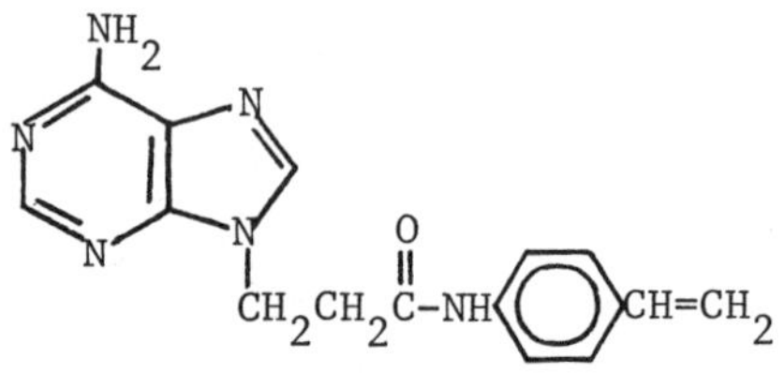

(XV)

Monomers (VII) and (VIII) have been studied the most extensively. The original work on polymerizing (VII) lead to an intractable, insoluble, cyclopolymerized polymer (40) but subsequent work lead to linear, water soluble polymers (41,42). Monomer (VIII) readily polymerizes under free radical conditions to form linear, water soluble polymers (27,43,44). The polymers of (VII) and (VIII) form complexes with each other in a manner similar to the natural nucleic acids (25,26) and they do show antitumor activity against murine leukemia virus (45,46).

Monomers (XI) and (XII) polymerize readily under free radical conditions to yield uncrosslinked polymers (31-36). The polymer from (XIb) was found to inhibit P388 leukemia in mice but it was uncertain whether this was due to activity of the polymer or to a slow release of 5-fluorouracil since this polymer (and monomer) do slowly hydrolyze in an aqueous medium (34). Monomers (IX), (X), (XIII), (XIV) and (XV) also polymerize but no biological data have been reported.

The first 5-fluorouracil containing polymer had the 5-FU units in the polymer backbone and this structure is shown below as (XVI). This was claimed to be biologically active (47). More recently, monomer (XVII) was prepared by the reaction of 5-FU with methyl fumaroyl chloride. The polymers and copolymers of (XVII) do show antitumor activity but this may be due to the hydrolysis of this unit to release 5-FU (48).

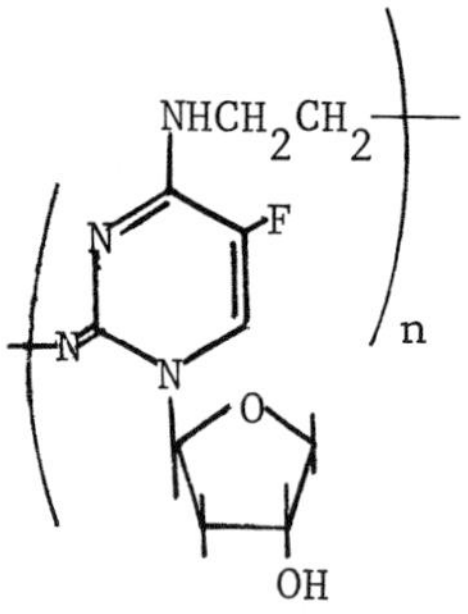

(XVI) 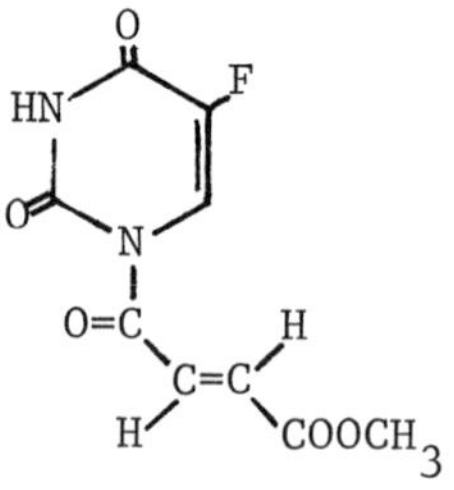   (XVII)

The first reported example of a polymer containing a 6-mercaptopurine derivative involved attaching the methylthio derivative to poly(vinyl alcohol) as shown in (XVIII). This water soluble polymer does form complexes with uracil containing polymers but no biological data was reported (49). Many other polymerizable derivatives of nucleic bases have been prepared including amino acid (50) and NCA derivatives (51) and derivatives of thymine, hypoxanthine and theophylline (8,25) but no information regarding possible antineoplastic activity is available.

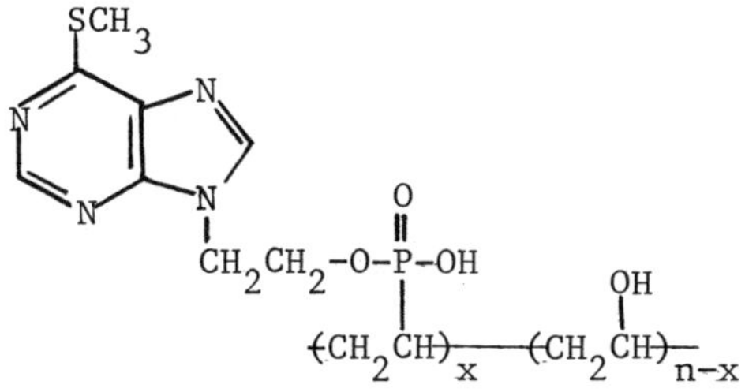

(XVIII)

It is obvious that this approach using nucleic bases can lead to antineoplastic polymers since several such systems have shown this activity (34,45,46,48). This is not totally unexpected since the nucleic base derivatives 5-fluorouracil and 6-mercaptopurine are antineoplastic agents. The polymers of this type made to date have been either the insoluble type or the soluble type polymeric drug systems. This approach should be especially effective if some directing group is also put into these polymers to carry them selectively into tumorous tissue. Research along this line is in progress in our laboratories and some other places as well.

Other Antineoplastic Polymers

Many other polymers with antineoplastic properties have been prepared. For the most part, these polymers contain a known antineoplastic compound attached to the polymer chain as a pendant group. The earliest examples of this type seem to be the aziridine alkylating group containing polymers (XIX) (52), and (XX) (53), and (XXI) (54) which contains the cytostatic agent sarcolysin.

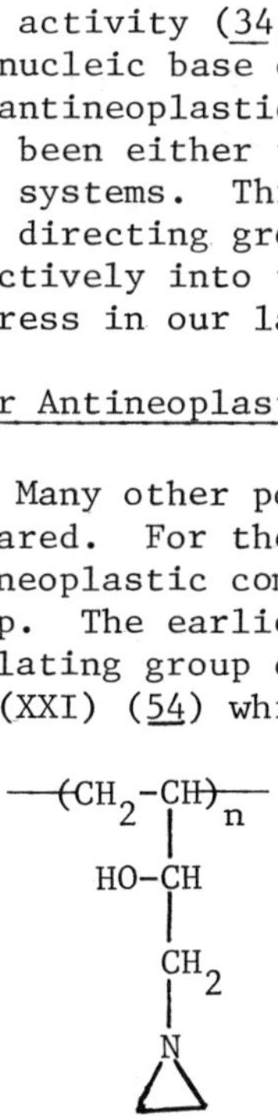

(XIX)

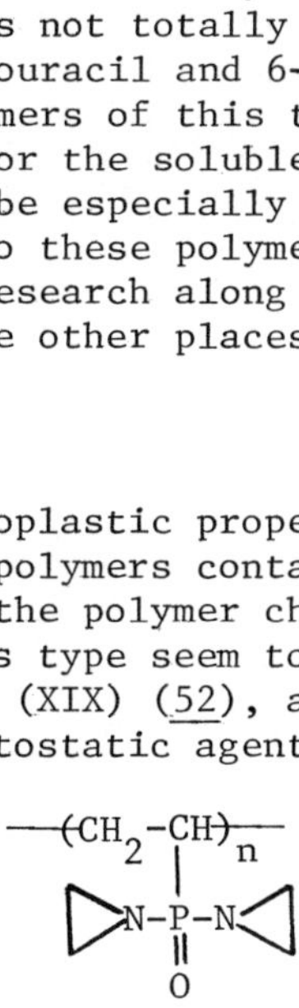

(XX)

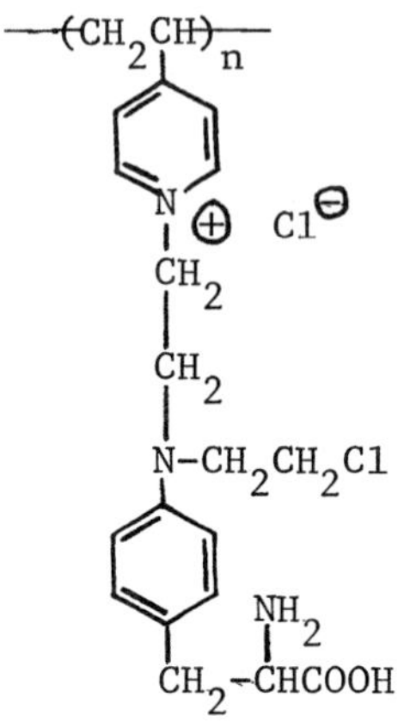

(XXI)

Other potential anticancer polymers also include the tropene derivative (XXII) (55,56) and the derivative of cyclophosphoramide (XXIII) (9,57,58). Monomer (XXIII) has also been copolymerized with solubilizing groups, such as N-vinylpyrrolidone or vinyl-pyridine-N-oxide, and some potential directing groups, which were polymerizable derivatives of sulfa drugs, to attempt to make a directed polymeric antineoplastic agent. This directed polymeric drug was initially claimed to migrate selectively into tumorous tissue (9).

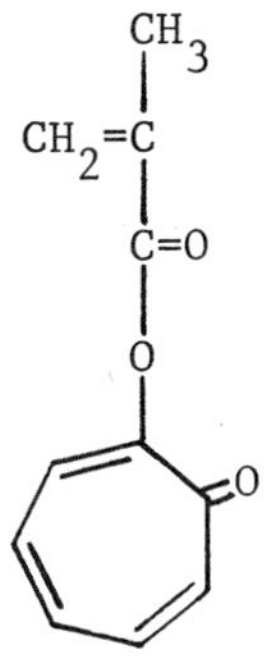

(XXII) 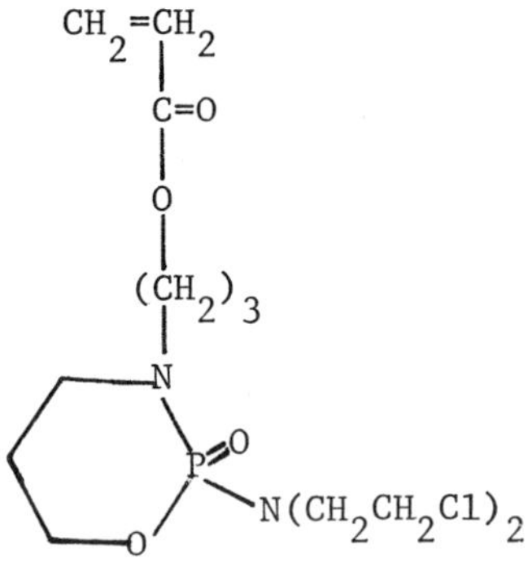   (XXIII)

The antineoplastic agents methotrexate and adriamycin have been bound to various protein substrates (59) and to DNA (60) and longer duration of the biological activity was noted compared to the monomeric drugs. The concept of affinity polymeric drugs, in which the drug unit is attached to some specific polymeric entity with affinity for some particular part of the body, has been reviewed recently (61). Polymeric derivatives of the anticancer agent cis-dichlorodiammineplatinium II have been prepared having the structure (XXIV) (62,63) and (XXV) (64). Polymer (XXIV) was found to be active against mouse P388 leukemia and Ehrlich ascites tumors (65) while polymer (XXV) was active against mouse 929 tumor cells and HeLa tumor cells (64).

(XXIV)

(XXV)

Conclusions

The present, brief review clearly shows that several different types of polymers can exhibit antineoplastic activity. It is not possible, at the present time, to state with certainty which polymeric system(s) will eventually prove to be the most useful in cancer therapy, but a directed polymeric antineoplastic agent would probably be the most desirable, if such a system can actually be synthesized. Some of the areas highlighted in this review will be considered further in some of the subsequent chapters of this book.

Literature Cited

1. Gebelein, C.G.; Morgan, R.M.; Glowacky, R.; Baig, W. "Biomedical and Dental Applications of Polymers"; Gebelein, C.G.; Koblitz, F.F., Ed.; Plenum Press: New York, 1981.
2. Samour, C.M. Chemtech 1978, 8, 494.
3. Gebelein, C.G. Polymer News 1978, 4, 163.
4. Donaruma, L.G.; Vogl, O. "Polymeric Drugs"; Academic Press: New York, 1977.
5. Batz, H.G. Adv. Polymer Sci. 1977, 23, 25; Springer-Verlag: New York.
6. Breslow, D.S. Pure & Appl. Chem 1976, 46, 103.
7. Kropachev, V.A. Pure & Appl. Chem 1976, 48, 355.
8. Takemoto, K. J. Polymer Sci 1976, Symp. 55, 105.
9. Ringsdorf, H. J. Polymer Sci 1975, Symp. 51, 135.
10. Donaruma, L.G. "Progress in Polymer Science"; Vol. 4, Jenkins, A.D., Ed.; Pergamon Press: New York, 1975; p. 1.
11. Yolles, S. "Polymers in Medicine and Surgery"; Kronenthal, R.L.; Oser, Z.; Martin, E., Ed.; Plenum Press: New York, 1975; p. 245.

12. Cowsar, D.R. "Polymers in Medicine and Surgery"; ibid., p. 237.
13. Zaffaroni, A; Bonsen, P. in Ref. 4, p. 1.
14. Zaffaroni, A. Chemtech 1976, 6 756.
15. Yoshida, M.; Kumakura, M.; Kaetsu, I. Polymer J. 1979, 11, 775.
16. Ottenbrite, R.M.; Regelson, W.; Kaplan, A.; Carchman, R.; Morahan, P.; Munson, A. in Ref. 4, p. 263.
17. Chem & Eng. News, Nov. 15, 1971, p. 79.
18. Regelson, W.; Holland, J.F. Nature (London) 1958, 181, 46.
19. Regelson, W.; Kuhar, S.; Tunis, M.; Fields, J.; Johnson, J.; Gluesenkamp, E. Nature (London) 1960, 186, 778.
20. Butler, G.B. J. Polymer Sci. 1960, 48, 279.
21. Solomon, D.H. J. Polymer Sci. 1975, Symp. 49, 175.
22. Levy, H.B. in Ref. 4, p. 305.
23. Morahan, P.S.; Munson, J.A.; Baird, L.G.; Kaplan, A.M.; Regelson, W. Cancer Res. 1974, 34, 506.
24. Morahan, P.S.; Kaplan, A.M. Int. J. Cancer 1976, 17, 82.
25. Takemoto, K. in Ref. 4, p. 103.
26. Pitha, J. Polymer 1977, 18, 425.
27. Ueda, N.; Kondo, K.; Kono, M.; Takemoto, K.; Imoto, M. Makromol. Chem. 1968, 120, 13.
28. Kaye, H. Polymer Letters 1969, 7, 1.
29. Kondo, K.; Iwasaki, H.; Ueda, N.; Takemoto, K.; Imoto, M. Makromol. Chem. 1968, 120, 21.
30. Akashi, M.; Kita, Y.; Inaki, Y.; Takemoto, K. Makromol. Chem. 1977, 178, 1211.
31. Gebelein, C.G.; Morgan, R.M. Polymer Preprints 1977, 18 (1), 811.
32. Gebelein, C.G.; Morgan, R.M.; Glowacky, R. Polymer Preprints 1977, 18 (2), 513.
33. Gebelein, C.G.; Ryan, T.M. Polymer Preprints 1978, 19 (1), 538.
34. Gebelein, C.G. Org. Coatings & Plastics Chem. 1980, 42, 422.
35. Gebelein. C.G.; Glowacky, R. Polymer Preprints 1977, 18 (1), 806.
36. Gebelein, C.G.; Baig, M.W. Polymer Preprints 1978, 19 (1), 543.
37. Hoffman, S.; Witkowski, W.; Schubert, H. Z. Chem. 1974, 14, 14.
38. Kondo, K.; Sato, T.; Inaki, Y.; Takemoto, K. Makromol. Chem. 1975, 176, 3505.
39. Kondo, K.; Ohbe, Y.; Takemoto, K. Makromol. Chem. 1976, 177, 3461.
40. Kaye, H. Macromolecules 1971, 4, 147.
41. Pitha, J.; Pitha, P.M.; Tso, P.O.P. Biochim. Biophys. Acta 1970, 204, 39.
42. Kaye, H.; Chang, S.H. Macromolecules 1972, 5, 397.
43. Pitha, J.; Pitha, P.M.; Stuart, E. Biochem. 1971, 10, 4595.

44. Kaye, H. J. Am. Chem. Soc. 1970, 92, 5777.
45. Pitha, P.M.; Teich, N.M.; Lowy, D.R.; Pitha, J. Proc. Nat. Acad. Sci., USA 1973, 70, 1204.
46. Vengris, V.E.; Pitha, P.M.; Sensenbrenner, L.L.; Pitha, J. Mol. Pharmacol. 1978, 14, 271.
47. Ballweg, H.; Schmael, D.; von Wedelstaedt, E. Arzneim. Forsch. 1969, 19 (8), 1296.
48. Umrigar, P.P.; Ohashi, S.; Butler, G.B. J. Polymer Sci., Chem Ed. 1974, 17, 351.
49. Seita, T.; Kinoshita, M.; Imoto, M. J. Macromol. Sci.-Chem. 1973, A7, 1297.
50. Doel, M.T.; Jones, A.S.; Taylor, N. Tetrahedron Letters 1969, 2285.
51. Takemoto, K.; Tahara, H.; Yamada, A.; Inaki, Y.; Ueda, N. Makromol. Chem. 1973, 169, 327.
52. Lidak, M.J.; Gillew, S.A. "Synthesis and Investigations of Cancerostatic Preparations"; State Publ. Med.: Moscow, 1962.
53. Nadzhimutdinov, S.; Kargin, V.A.; Usmanov, K.U.; Bruevich, G.Y.; USSR Patent 260,887, 1970; C.A. 1970, 72, 133.
54. Zubova, O.V.; Kirsh, Yu.E.; Lebedeva, T.S.; Shorokhova, A.A.; Silaev, A.B.; Kabanov, V.A.; Kargin, V.A. Dokl. Akad. Nauk., SSSR 1969, 186, 477.
55. Cornell, R.J.; Donaruma, L.G. J. Med. Chem. 1965, 8, 388.
56. Cornell, R.J.; Donaruma, L.G. J. Polymer Sci., Pt. A 1965, 3, 827.
57. Bartulin, J.; Przybylski, M.; Ringsdorf, H.; Ritter, H. Makromol. Chem. 1974, 175, 1007.
58. Batz, H.G.; Ringsdorf, H.; Ritter, H. Makromol. Chem. 1974, 175, 2229.
59. Whiteley, J.M.; Chu, B.C.F.; Galivan, J. in Ref. 1, p. 241.
60. Chu, B.C.F.; Whiteley, J.M. Mol. Pharmacol. 1977, 13, 80.
61. Goldberg, E. in Ref. 4, p. 239.
62. Allcock, H.; Allen, R.; O'Brien, J. J. Chem. Soc., Chem. Comm. 1976, 717.
63. Allcock, H.; Cook, W.; Mack, D. Inorg. Chem. 1972, 4, 2584.
64. Carraher, C.E., Jr. in Ref. 1, p. 215.
65. Allcock, H. "Organometallic Polymers"; Carraher, C.; Sheats, J.; Pittman, C., Ed; Academic Press: New York, 1978; p. 283.

RECEIVED August 12, 1981.

16

The Antitumor and Antiviral Effects of Polycarboxylic Acid Polymers

RAPHAEL M. OTTENBRITE

Virginia Commonwealth University, Department of Chemistry, Richmond, VA 23284 and Medical College of Virginia Cancer Center, Virginia Commonwealth University, Richmond, VA 23298

Anionic and cationic polyelectrolytes of both natural and synthetic origin have been found to exhibit an inhibitory effect on viruses, bacteria, tumors, and enzymes (1). Polyanions, in particular, have a broad range of biological activity and have received considerable interest in the areas of oncology and virology. The prolonged protective action of synthetic polyanions when given prior to virus inoculation has tremendous clinical potential; consequently, establishing an impetus for assaying the fundamental role of polyanions in controlling host resistance to a variety of pathophysiology.

Synthetic polyanions, in particular, are known to produce a wide spectrum of effects on immune reactivity (1). They have been shown to induce production of interferon (2) modify RES function (3) and to have immunoadjuvant (4) antiviral (5) and antitumor activity (6). The antineoplastic activity of pyran copolymer has been largely attributed to its ability to activate macrophages (7).

One of the most interesting anionic polymers is pyran copolymer which is prepared from divinyl ether and maleic anhydride through copolymerization. It has been under investigation in cancer chemotherapy for several years and is designated as NSC 46015 by the National Cancer Institute. It shows a variety of biological activities and has elicited a considerable amount of interest by several researchers in different areas. Pyran copolymer is an inducer of interferon (8-11); it has activity against a number of viruses (6-14), including Friend leukemia, Rauscher leukemia, Moloney sarcoma, polyoma, vesicular stomatitis, Mengo, encephalomyocarditis, MM, and foot-and-mouth disease; it has antibacterial (21-23) and antifungal activity (21); it stimulates immune response (18-24); it inhibits adjuvant disease (31); it is an interesting anticoagulant (32); and it shows promise in removing plutonium from the liver (33). However, the toxicity of the copolymer, although much lower than that of other anionic polymers which have been investigated, was apparently still too high for it to be used extensively in clinical investigations

(34,35). Recent studies have shown that low molecular weight fractions and the calcium salt of the polymer are much less toxic. This has stimulated further clinical investigation of this drug.

Antitumor and Antiviral Activity of Polycarboxylic Acid Polymers

We have found that the structure and the molecular weight of the polycarboxylic acid polymers play an important role in the biological activity elicited by these materials (1). These findings have been coorborated by Breslow (36) who has reported that fractionated pyran shows lower toxicity with lower molecular weight fractions with very little variation in antitumor activity. Hodnett (37) has also prepared a number of polycarboxylic acid copolymers which gave varying activity with structure against ascites sarcoma 180 tumor. The average increase in survival time was 25%. He found that the most effective polymers were those with low molecular weight, high carboxyl group density and low pKa at physiological pH.

We have prepared several different molecular fractions of pyran, poly(acrylic acid-co-maleic acid) PAAMA, poly(maleic acid) PMA, and poly(acrylic acid-co-3,6-endoxo-1,2,3,6-tetrahydrophthalic acid) BCEP and evaluated their activity against Lewis lung carcinoma (Table I) and encephalomyocarditis (EMC) virus (Table II).

Table I. Inhibition of Tumor Size with Polymer Molecular Weight[a]

POLYMER[b]	PYRAN (%)	PAAMA (%)	PMA (%)	BCEP (%)
Whole Polymer	78	74	75	30
(Increased life span)	(40)	(33)	(15)	(<10)
1,000-10,000 (A)[c]	89	–	74	22
10,000-30,000 (B)	84	80	72	–
30,000-50,000 (C)	74	–	76	–
50,000-100,000 (D)	70	78	74	30

[a] BDF, mice were inoculated with 10^6 Lewis' lung cells into right-hind gluteus muscle. Polymers were administered daily by the intraperitoneal route for 10 consecutive days following tumor implantation. Primary tumor size was determined on day 14 (% inhibition) and increased life span (% ILS) calculated from mean time to death.

[b] Poly(divinylether-co-maleic anhydride) (Pyran); Poly(acrylic acid-co-maleic anhydride) (PAAMA); Poly(maleic anhydride) (MA); Poly(

acrylic acid-co-3,6-endoxo-1,2,3,6-tetrahydrophthalic anhydride) (BCEP).

cFRACTION A = PM-10 Filtrate passed through Amicon PM-10 filter and not UM-2
FRACTION B = PM-30 Filtrate passed through Amicon PM-30 filter and not PM-10
FRACTION C = XM-50 Filtrate passed through Amicon XM-50 filter and not UM-2
FRACTION D = Whole Polymer passed through Amicon XM-100 filter and not XM-50

Table II. Change in Antiviral Encephalomyocarditis Protection with Polymer Molecular Weight^d

POLYMER		PYRAN (%)	PAAMA (%)	PMA (%)	BCEP (%)
Whole Polymer		89	90	30	25
A	1,000-10,000	0	0	0	0
B	10,000-30,000	30	29	0	-
C	30,000-50,000	58	54	<10	-
D	50,000-100,000	86	80	26	38

^{d}Mice were inoculated intravenously with 10 LD_{50} of encephalomyocarditis virus 24 hours after administration (i.v.) of 25 mg/kg of polymer. Percent protection based on 20 mice/group calculated from mean time to death.

Polymer structure changes showed different effects in tumor size. All the polymers reduced the tumor growth by approximately 75% except for BCEP which only inhibited the growth by 30%. More importantly each polymer had a different effect on the increased life span (ILS) of the animal with pyran being the most effective (40% ILS) and BCEP the least (10% ILS). The molecular weight did not have significant effect on the antitumor activity but did have a great effect on toxicity of the material which will be discussed later.

The polymer structure had a more significant effect on antiviral activity (Table II). Pyran and PAAMA produced approximately the same activity while PMA and BCEP were very poor. Further, molecular weight had a dramatic effect. The lower molecular weight fractions (<30,000) were ineffective and only the higher molecular weight materials (>30,000) were active.

In a more recent study by Munson et. al. (38), it was shown that pyran (Table III) elicited better cytotoxicity against Lewis lung with the lower molecular weight fractions. It was also

Table III. Relationship Between Molecular Weight, Macrophage Activation, Antitumor Activity and Antiviral Activity

MW Pyran Polymer	Percent cytotoxicity[e]		Antitumor[f] Activity P815 mastocytoma % ILS	mg/kg Pyran[g] Required for 100% Protection Against EMC
	Lewis lung	mouse embryo fibroblast		
12,500	100	0	126	<75
15,500	94	0	126	25
21,300	94	0	179	12.5
32,000	81	0	179	6.3
52,600	88	0	162	3.1
NSC46015 (pyran)	88	0	not done	3.1

[e] Peritoneal exudate cells were removed and tested 7 days after i.p. injection of pyran polymers at 25 mg/kg for cytotoxicity against Lewis lung and secondary mouse embryo cells.

[f] Mice were inoculated in the footpad with 5×10^5 P815 mastocytoma cells.

[g] Mice were injected i.v. with pyran polymers and challenged i.v. 24 hours later with 10 LD_{50} of EMC virus.

shown in this study that there was no cytotoxicity of the natural mouse embryo fibroblast with any pyran; thus the activity elicited by pyran was tumor cell specific. Interestingly, increased life span was observed for the high molecular weight pyran in treatment against P815 mastocytoma. Furthermore, much less higher molecular weight (52,600) pyran (3.1 mg/kg) was required to produce 100% protection against EMC whereas it required 75 mg/kg of the 12,500 molecular fraction to produce the same result.

More recently, we have evaluated a number of maleic anhydride copolymers (Table IV) and 2,3-dicarboxynorborn-5-ene copolymers (Table V) against Ehrlich ascites tumor (33). In both of these systems, very little or no activity was observed for the 10,000-30,000 molecular weight fraction compared to pyran. However, the lower molecular weight fractions (1,000-10,000) all produced good activity and in many cases were better than pyran.

Table IV. Effect of Maleic Anhydride Copolymers Against Ehrlich Ascites Tumor Cells[h]

POLYMER	10,000 - 30,000 Molecular Weight		1,000 - 10,000 Molecular Weight	
	Cells/pc x 10^8	% Control[i]	Cells/pc x 10^8	% Control
Saline	1.52	100	2.17	100
Pyran	0.74	49	.63	30
MA-co-Acrylic Acid	1.8	118	1.25	58
MA-co-STYRENE	0.76	45	0.15	7
MA-co-Methracylic Acid	1.42	93	1.31	60
MA-co-Allyl Phenol	1.31	86	0.35	16
MA-co-Allyl Succinic Anhydride	1.32	87	0.91	42
MA-co-1,3-Dioxepin	1.43	94	0.63	30
MA-co-Isobutenyl Succinic Anhydride	1.21	80	1.33	61

h) 10^6 Ehrlich Ascites cell were inoculated i.p. into mice.

i) Five days after inoculation, animals were sacrificed and total peritoneal exudate cells were counted with a hemocytometer.

Table V. Effect of 2,3-Dicarboxynorborn-5-ene Copolymers Against Ehrlich Ascites Tumor[j]

COPOLYMER	10,000 - 30,000 Molecular Weight		1,000 - 10,000 Molecular Weitht	
	Cells/pc x 10^8	% Control[k]	Cells/pc x 10^8	% Control
Control	1.52	100	2.17	100
Pyran	0.74	49	.64	30
Maleic Anhydride	1.64	1.08	0.16	7
Acylic Acid	1.61	1.06	0.48	22
Vinyl Acetate	0.89	41	0.76	34
Vinyl Alcohol	0.85	56	0.71	32

j) 10^6 Ehrlich Ascites cells were inoculated i.p. into mice.

k) Five days after inoculation, animals were sacrificed and total peritoneal exudate cells were counted with a hemocytometer.

Consequently, we are observing two significant effects with regard to antitumor and antiviral activity. These effects appear to be related in most cases, to the polymer structure and molecular weight.

Toxicological Effects of Polycarboxylic Acid Polymers

The toxicity of polycarboxylic acid polymers is more sensitive to polymer structure and molecular weight than antitumor activity. The acute toxicity is markedly decreased with decreasing molecular weight (Table VI and XI). Further, polyanion polymers have been shown to highly sensitize mice to bacterial endotoxin (39); this activity is strikingly molecular weight dependent as shown in Tables VII and XI as well as dose dependent (38). The enhanced sensitivity to endotoxin by synthetic anionic polymers has been extensively studied, but the mechanism of action is still not understood (39).

Table VI. Polymer Molecular Weight Effect on Acute Toxicity $(LD_{50})^1$

POLYMER		PYRAN	PAAMA	PMA	BCEP
Whole Polymer		74	110	120	150
A	1,000-10,000	120		160	>200
B	10,000-30,000	115	>200	150	-
C	30,000-50,000	95		140	-
D	50,000-100,000	84	180	135	180

1) LD_{50} was determined by administering the polymer in mg/kg intraveneously and the mortality recorded 24 hours later.

Table VII. Effect of Endotoxin Sensitization With Polymer Molecular Weight[m]

POLYMER		PYRAN	PAAMA	PMA	BCEP
Whole Polymer		0.12	1.0	15	3.0
A	1,000-10,000	15	-	>20	>10
B	10,000-30,000	15	-	>20	>10
C	30,000-50,000	-	>20	>20	>3
D	50,000-100,000	-	>20	>20	>3.0

m) LD_{50} of S. typhosa 0904 lipopolysaccharide 24 hours after a single administration of 24 mg/kg of polymer. The dosage of endotoxin was administered in mg/kg.

Molecular weight toxicities are also observed in the liver (Tables VIII and IX) as well as in the spleen as shown in Table X (1). These effects are also polymer structure and molecular weight dependent. The larger molecular weight polymers seem to be localized in the liver and in the spleen to a greater extent than the lower molecular weight polymers; which appear to affect the reticular and other cells of the spleen as well as the Kupffer cells, mixed microsomial enzymes and other factors in the liver (1).

In a recent study by Munson, White and Klykken (38), it was reported that the acute toxicity of pyran was a function of salt form. It was demonstrated that the calcium salts were less toxic than the sodium salts of pyran and that the toxicity in both cases was molecular weight dependent (Table XI) as well as dose dependent (38). In contrast to the significant differences in acute toxicity observed for the two salts, no significant effects were seen on the sensitization to bacterial endotoxin (Table XI) clinical chemistry, or antiviral activity (38).

Table VIII. Polymer Molecular Weight Effect on Liver Weight[n]

POLYMER		PYRAN	PAAMA	PMA	BCEP
Whole Polymer		7.8	6.9	7.4	6.4
A	1,000-10,000	5.1	-	7.1	5.8
B	10,000-30,000	5.9	-	6.8	5.9
C	30,000-50,000	-	4.9	6.8	6.2
D	50,000-100,000	-	5.4	7.3	6.3

n) Polymers administered in a dose of 25 mg/kg intravenously. Organ weights determined seven days after drug injection and expressed as percent of total weight. Liver size of control group was 5.4.

Table IX. Hexobarbitol Sleeping Time Effected by Polymer Molecular Weight[o]

POLYMER		PYRAN	PAAMA	PMA	BCEP
Whole Polymer		98	64	58	62
A	1,000-10,000	43	-	44	38
B	10,000-30,000	47	-	46	38
C	30,000-50,000	-	41	44	40
D	50,000-100,000	-	48	49	44

o) Mice were inoculated with 25 mg/kg i.v. of the polymer. Twenty four hours later an anesthetic dose (80 mg/kg) of sodium hexobarbital was administered i.v. and duration of anesthesia recorded in minutes. Sleeping time of control group was 37 minutes.

Table X. Polymer Molecular Weight Effect on Spleen Weight[p]

POLYMER		PYRAN	PAAMA	PMA	BCEP
Whole Polymer		1.08	0.98	.90	0.84
A	1,000–10,000	0.40	–	.60	0.78
B	10,000–30,000	0.44	–	.73	0.78
C	30,000–50,000	–	.41	.84	0.82
D	50,000–100,000	–	.58	1.04	0.83

p) Polymers administered in a dose of 25 mg/kg intraveneously. Organ weights determined 7 days after drug injection and expressed as percent of total body weight. Spleen size of control group was 0.36.

Table XI. Effect of Molecular Weight of Pyran on Acute Toxicity and Sensitivity to Bacterial Endotoxin.

POLYMER MW	LD_{50}[q]		LD_{50} of Bacterial[r] Endotoxin	
	Na^+ Salt	Ca^{++} Salt	Na^+ Salt	Ca^{++} Salt
12,500	112	–	24.0	–
15,500	98	190	7.0	10
21,300	94	–	0.8	–
52,600	86	170	0.5	0.5
NSC-46015	78	–	0.3	–
Control	–	–	21.0	21.0

q) Mice were injected i.v. with sodium and calcium salts of Pyran. Mortality was observed over a 14 day period.

r) Mice were injected i.v. with 25 mg/kg of pyran followed 24 hours later by i.s. administration of S. typhosa lysopolysaccharide 0.901. Control vehicles were NaCl and bacterial dosage were in mg/kg.

Macrophage Activation by Polycarboxyl Acid Polymer Immunopotentiators

Macrophages are produced by rapidly dividing precursor cells in the bone marrow and the daughter cells are released into the blood-stream. This mononuclear phagocytic system consists of several types of cells widely distributed throughout the body. Once they leave the blood, the macrophage settle in the tissues of the liver, spleen, lymph nodes, lungs, and peritoneum. In these sites, they differentiate into the various types of fixed macrophages (40,41).

Initially, macrophages were considered as the "antigenic garbage disposal unit," concerned only with the phagocytosis and the degradation of antigen. In recent years, however, macrophages have undergone considerable reevaluation with regard to their role in the immune system. The function that macrophage play in the generation of the humoral response has been extensively studied (40) and postulated mechanisms of macrophage activity in antibody production include; antigen processing for "presentation" to the lymphocytes and antigen processing by elaboration of soluble factors. Wagner et al. (42) have also reported that the expression of cell-mediated immunity is primarily dependent on the interaction between T-lymphocytes and macrophages.

It has been observed that under suitable conditions, macrophages can transform into "activated" macrophages, characterized by enlarged cells with undulating (ruffled) plasma membranes. Several lines of evidence indicate that activated macrophages are both qualitatively and quantitatively different from normal macrophages.

Although several biologic reticuloendothelial stimulants such as Mycorbacilerium bovis, bacille Calumette-Guerin, C. parvum and Toxoplasma gondii, are known to enhance macrophage function as well as to induce resistance to tumor growth (43), synthetic polyanionic immunopotentiators such as pyran and polyacrylic acid have been shown to be potent activators as well. Macrophage treated with these polyanionic polymer stimulants have shown cytotoxicity for normal cells (44). It has been demonstrated in our laboratories at MCV/VCU that the immunopotentiator pyran causes an increase in the number of histocytes present in the connective tissue surrounding the Lewis lung tumor of mice (45). In addition, increased numbers of tumor-associated macrophages are correlated with a decrease in metastasis in pyran-treated tumor-bearing mice (46).

It is not clear whether cytotoxicity and cytostasis involve the same function of the macrophages. Cytostasis by pyran-activated macrophages, similar to other stimulants, occurs as early as 4-6 hours after incubation of target cells with the macrophages (47). Although the growth of tumor cells is inhibited, ^{125}IUDR is not released from target cells until 48 hours after initation (48). Similar behavior has also been reported for macrophages activated by other agents as well (49). The possible involvement of "factors" produced by activated macrophages in the tumorcidal event is receiving considerable attention (50-52).

Recent data which has been reported indicates that macrophages activated by pyran increase biochemical activity, are cytotoxic for tumor cells, and have a new macrophage cell surface antigen (AMØSCA). In contrast, this new macrophage cell surface antigen was not detected on normal macrophages or macrophages elicited by glycogen or thioglycollate, which are known to have increased biochemical activity but are not cytotoxic for tumor cells (53). AMØCSA can be detected within six hours after i.p. inoculation with pyran and its appearance correlated with the progression of macrophages to cytotoxic effector cells (53). The excellent correlation of AMØCSA with cytotoxicity and in vivo antitumor activity provides us with a simple, rapid technique for analyzing macrophage activation. Moreover, since polyanions such as pyran and dextran sulfate have been shown to activate macrophages in vitro (54) this provides a model system for analyzing the chemical moiety associated with these agents which is responsible for anti-tumor activity. Since their ability to activate macrophages has been associated with the anti-tumor activity of these agents this approach could provide a powerful tool for quickly screening potentially active compounds.

The cytotoxicity of activated macrophage has been demonstrated to be confined to only tumor cells (55-58) and has an insignificant or no effect on normal or natural cells such as mouse embryo cells (59) or murine kidney cells (56,58). The mechanism of activation of macrophage or the tumoricidal activity of activated macrophage is not understood. It has been observed that a population of aneuploid Lewis lung carcinoma cells, in the presence of activated macrophage, show reductive antitotic cell division resulting from tumor cells with 50% less DNA per cell (59). This depletion of DNA leads to cell destruction since these cells no longer have the appropriate DNA to synthesize necessary molecules for cell division.

In Table XII, cell cycle analysis shows (60) that in vitro Lewis lung cells are distributed such that 15% are in the resting phase (G_1) 52% in the DNA synthesis phase (S) and 33% are in cell division phases (G_2M). Following 24 hours inculation with pyran activated macrophage, the percent tumor cells in the resting phase (G_1) rose to 74% while synthesis (S) and cell division phases fell to 22% and 4% respectively. The shift to G_1 phase was not seen with Lewis lung cells cultivated with normal macrophage or mouse embryo fibroblasts incubated with activated macrophage.

Table XII. DNA Distribution of Lewis Lung Cell Cultivated With Pyran Activated Macrophages

Lewis Lung Cultured[s] With AMØ for	Percent in Phase[t]		
	G_1	S	G_2M
0 hr	15	52	33
4 hr	16	52	32
8 hr	32	47	20
16 hr	62	34	4
24 hr	74	22	4

s) Activated macrophage (AMØ) were obtained from mice injected i.p. with 25 mg/kg of pyran 7 days before peritoneal lavage.

t) A pattern recognition program was used to determine G_1, S and G_2M phase.

Animals treated i.p. with pyran produced a 30 fold inhibition of Ehrlich ascites tumor growth with maximum inhibition of tumor growth occuring with pyran treatment two days prior to tumor cell inoculation (Table XIII) (60).

Similar to the in vitro study using Lewis lung cells, it was observed that animals pretreated with pyran had no detectable Ehrlich ascites cells in phase G_2M and only a few in the S phase after 2 or 6 days (Table XIV) (60) compared to pyran untreated mice which had 58% in the S phase and 22 in the G_2M phase. Consequently, the use of pyran activated macrophage resulted in a) a high overall level of tumoricidal activity both in vivo and in vitro, b) the appearance of a tumor cell population with 50% of their normal DNA content and c) a shift of tumor cells from phase G_2M to G_1. These results indicate that a possible mechanism of activated macrophage tumoricidal activity involves the induction of tumor cells with a reduced DNA content.

Table XIII. Inhibition of Ehrlich Ascites Tumor Cell Proliferation in vivo by Pyran Inoculation

Macrophage Inducer	Number of Days Between Inoculation of Macrophage Pyran[u] and EA Cells	Total PEC Counts Per Animal $(\times 10^{-7})$[v]
Pyran	2	1.6
Pyran	5	18
Pyran	7	28
Tumor Control	-	35

u) At various days, prior to tumor cell inoculation, animals received i.p. inoculation of pyran (25 mg/kg). 1×10^6 EA cells were inoculated i.p. into mice on day zero.

v) Five days after tumor inoculation, animals were killed and the total peritoneal exudate cells (PEC) counted.

Table XIV. Cell Cycle Analysis of Ehrlich Ascites Tumor Cells Before and After in vivo Exposure to Pyran

Days Following Tumor Inoculation[w]	Pyran Treatment[x]	Percent in Phase[y]		
		G_1	S	G_2M
2	-	20.	58	22
2	+	98	2	0
6	-	36	47	17
6	+	98.	2.	0

w) Mice were inoculated i.p. with 1×10^6 EA cells and killed on various days thereafter.

x) Pyran was injected i.p. (23 mg/kg) 2 days prior to tumor inoculation.

y) Cell cycle analysis of peritoneal EA cells was determined with a cell sorter utilizing a pattern recognition program.

Clinical Effects of Polycarboxylic Acid Polymers

Most polyanions are water soluble, which is essential for biological transport and more importantly for systemic administration, since injection of suspensions into blood system can cause "colloido-clasmic shock" or "macromolecular syndrome" with major hypersensitive clinical toxicity. Water-soluble polyanionic polymers can be distributed in a living system by means of blood or lymphatic circulation, by cellular transport through the involvement of mobile phagocytic cells, and absorption on cell surfaces.

In the initial study, pyran copolymer was given to advanced cancer patients who were no longer responsive to other treatment modalities. Survival had to be estimated at one month and patients had to be off other forms of chemotherapy or radiotherapy at least two weeks without signs of narrow depression or active sepsis and liver disease. Pyran copolymer was then given to 62 patients and the consistent limiting toxicity of pyran was transient thrombocytopenia. Cytoplasmic inclusions were seen in circulating leukocytes, nucleated marrow cells, and phagocytic cells of liver and spleen after administration. Although not dose related, fever was seen in 50% of patients.

The recent development of less toxic, lower MV fractions; less sensitive interpreitoneal or intervenous administration; as well as the formulation of the polyanion hydrolyrate as the more compatable calcium salt has stimulated renewed interest in at least two clinical studies now in progress. Future clinical studies should involve investigations of polyanions as immuno-adjuvants for antitumor and antiviral vaccines.

Acknowledgments

The authors wish to acknowledge the support in part of this research by NIH grant AI-15612-02 as well as P. Grover and A. Howell for their help in preparing this manuscript.

Literature Cited

1. Ottenbrite, R. M.; Donaruma, L. G.; Vogl, O. "Antionic Polymeric Drugs," John Wiley and Sons, N.Y. 1980.
2. Merigan, T. C.; Regelson, W. N. Eng. J. Med., 1967, 277: pp. 1283-1287.
3. Munson, A. E.; Regelson, W.; Lawrence, W.; Wooles, W. R. J. Reticuloendothel. Soc., 1970, 7: pp. 375-385.
4. Baird, L. G.; Kaplan, A. M. Cell Immunol., 1975, 20: pp. 167-176.
5. Morahan, P. S.; Regelson, W.; Munson, A. E. Antimicros. Ag. Chem., 1972, 2: pp. 16-22.
6. Morahan, P. S.; Kaplan, A. M. Int. J. Cancer, 1976, 17: pp. 82-89 and Mohr, S. J.; Chirigos, M. A.; Fuhrman, F. S.; Pryor, J. W. Cancer Res., 1975, 35: 3750-3654.

7. Morahan, P. S.; Kaplan, A. M. Int. J. Cancer, 1976, 17, pp. 82-89.
8. Merigan, T. C. Nature, 1967, 214, 416.
9. Merigan, T. C.; Regelson, W. N. Engl. J. Med., 1967, 277, 1283.
10. Merigan, T. C. Ciba Foundation Symposium on Interferon, ed., Wolstenholme, G. E. W. and O'Connor, M. J. & A. Churchill Ltd., London, 1967, pp. 50-60.
11. DeClercq, E.; Merigan, T. C. Arch. Intern. Med., 1970, 126, 94.
12. Regelson, W. Adv. Exp. Med. Biol., 1967, 1, 315.
13. Merigan, T. C.; Finkelstein, M. S. Virology, 1968, 35, 363.
14. DeClercq, E.; Merigan, T. C. J. Gen Virol., 1969, 5, 359.
15. Chirigos, M. A.; Turner, W.; Pearson, J.; Griffin, W. Int. J. Cancer, 1969, 4, 267.
16. Chirigos, M. A. Comparative Leukemia Research 1969. Bibl. Haemat., No. 36, ed., Dutcher, R. M., pp. 278-292 (Karger, Basel/Munchen/Paris/New York, 1970).
17. Schmidt, J. P.; Pindak, F. F.; Giron, D. J.; Ibarra, R. R. Texas Rep. Biol. Med., 1971, 29, 133.
18. Richmond, J. Y. Infec. Immun., 1971, 3, 249. Arch. ges. Virusforsch., 1972 36, 232.
19. Schuller, G. B.; Morahan, P. S.; Snodgrass, M. J. 10th National Meeting of the Reticulo. Soc., Abs. 1973, 28.
20. Campbell, C. H.; Richmond, J. Y. Infec. Immun., 1973, 7, 199.
21. Regelson, W.; Munson, A.; Wooles, W. Internat. Symp. on Stand. of Interferon and Interferon Inducers, London, 1969; Symp. Series Immunobiolog. Stand., Vol. 14, pp. 227-236 (Karger, Basel/N.Y., 1970).
22. Pindak, F. F. Infec. Immun., 1970, 1, 271.
23. Giron, D. J.; Schmidt, J. P.; Ball, R. J.; Pindak, F. F. Antimicrob. Agents and Chemother., 1972, 1, 80.
24. Regelson, W.; Munson, A. E. Ann. N.Y. Acad. Sci., 1970, 173, 831.
25. Munson, A. E.; Regelson, W.; Lawrence, W., Jr.; Wooles, W. R. J. Reticuloendothel. Soc., 1970, 7, 375.
26. Braun, W.; Regelson, W.; Yajima, Y.; Ishizuko, M. Proc. Soc. Exp. Biol. Med., 1970, 133, 181.
27. Hirsch, M. S.; Black, P. H.; Wood, M. L.; Monaco, A. P. Proc. Soc. Exp. Biol. Med., 1970, 134, 309.
28. Kapila, K.; Smith, C.; Rubin, A. A. J. Reticuloendothel. Soc., 1971, 9, 447.
29. Hirsch, M. S.; Black, P. H.; Wood, M. L.; Monaco, A. P. J. Immunol., 1972, 108, 1312.
30. Hirsch, M. S.; Black, P. H.; Wood, M. L.; Monaco, A. P. Personal communication.
31. Kapusta, M. A.; Mendelson, J. Arthritis Rheum., 1969, 12, 463.
32. Shamash, Y.; Alexander, B. Biochim. Biophys. Acta, 1969, 194, 449.

33. Rosenthal, M. W. Argonne National Laboratory, personal communication.
34. Leavitt, T. J.; Merigan, T. C.; Freeman, J. M. Am. J. Dis. Child., 1971, 121, 43.
35. Regelson, W. Personal communication.
36. Breslow, D. S. Pure and Appl. Chem., 1976, 46, pp. 103-113 and Polymer Preprints, 1981, 22, 24.
37. Hodnett, E.; Amirmoazzamo, J.; Tien Hal Tai, J. J. Med. Chem., 1978, 21, 652.
38. Munson, A. E.; White; Klykken, P. Cancer Res., 1981, 16, 329.
39. Munson, A. E.; Regelson, W. Proc. Soc. Exp. Biol. Med., 1971, 137, 553.
40. Abdour, N. I.; Richter, M. Adv. Immunol., 1970, 12, 202.
41. Unanue, E. R. Adv. Immunol., 1972, 15, 95.
42. **Wagner, H.; Feldmann, M.; Boyle, W.; Schrader, J. J. Exp. Med., 1971, 136, 331.**
43. **Holterman, O. A.; Klein, E.; Casale, G. P. Cell Immunol., 1973, 9, 339.**
44. Kaplan, A. M.; Morahan, P. S.; Regelson, W. J. Natl. Cancer Inst., 1974, 52, 1919.
45. Snodgrass, M. J.; Kaplan, A. M.; Morahan, P. S. Cancer Res., 1975, 55, 455.
46. Eccles, S. A.; Alexander, A. Nature, 1974, 250, 667.
47. Evans, R. In, Macrophage Activation, Wagner, W. H.; Hahn, H. Eds., Experpta Medica, Amsterdam, Netherlands, 1974, p. 305.
48. Kaplan, A. M.; Walker, P. L.; Morahan, P. S. In, Modulation of Host Immune Resistance, Chirigos, M. Ed., Fogarty International Center Proc., U. S. Government Printing Office, Washington, D. C., No. 28, 277, 1977.
49. Olivotto, M.; Bomford, R. Int. J. Cancer, 1974, 13, 478.
50. Currie, G. A.; Basham, C. J. Exp. Med., 1975, 142, 1600.
51. Melsom, H.
52. Meltzger, M. S.; Bartlett, G. L. J. Natl. Cancer Inst., 1972, 49, 1439.
53. Kaplan, A. M.; Mohanakumar, T. J. Exp. Med., 1977, 146, 1461.
54. Schultz, R. M.; Papamatheakis, J. D.; Chirigos, M. A. Cellular Immuno., 1977, 29, 403.
55. Poste, G.; Kirsh, R.; Fogler, W. E.; Tidler, I. J. Cancer Res., 1979, 39, 881.
56. Holteman, O. A.; Klien, E.; Casal, G. P. Cell. Immunol., 1973, 9, 339.
57. Russell, S. W.; Doe, W. F.; Cochrane, C. G. J. Immunol., 1976, 116, 164.
58. Currie, G. A.; Basham, C. Br. J. Cancer, 1978, 38, 653.
59. Melter, M. S.; Tucker, R. W.; Breuer, A. C. Cell. Immunol., 1975, 17, 30.
60. Kaplan, A. M.; Connolly, K. M.; Regelson, W., In "The Host Invader Interplay" ed., H. Van Vendbosshe, Elsevier/North Holland, Netherlands, p. 479, 1980.

RECEIVED August 26, 1981.

Polymeric Derivatives Based on cis-Diamminedichloroplatinum(II) as Antineoplastic Agents

CHARLES E. CARRAHER, JR., WILLIAM J. SCOTT, ISABEL LOPEZ, DELIE ROSELYN CERUTIS[1], and TUSHAR MANEK

Wright State University, Department of Chemistry, Dayton, OH 45435

DAVID J. GIRON

Wright State University, Department of Microbiology and Immunology, Dayton, OH 45435

Platinum polyamines can be readily synthesized with good control on the chain length through reaction of salts of PtX_4^{-2} with diamines. The

$$PtX_4^{-2} + H_2NRNH_2 \longrightarrow \left(Pt\begin{matrix}X & X\\ NH_2-R-NH_2\end{matrix}\right)$$

polyamines show good antineoplastic activity against a wide range of tumors including mouse connective tissues, human cervical carcinoma and human amnion cancer cells. Further the vast majority of the polyamines successfully altered the normal replication cycle of the polio virus Type 1 and Encephalomyocarditic virus, strain MM when the former cells were treated with the virus, without destruction of the cells themselves. Mice are able to tolerate large dosages of the polyamines.

Malignant neoplasms are the second leading cause of death in the United States. In 1964 Rosenberg and coworkers discovered

---

[1] Also affiliated with Wright State University, Department of Microbiology and Immunology, Dayton, OH 45435.

that bacteria failed to divide, but continued to grow giving filamentous cells in the presence of platinum electrodes (1). A major cause of this inhibition to cell division was cis-dichlorodiamineplatinum II,c-DDP. Much work centered about the clinical use of c-DDP leading to the licensing of it as an antineoplast drug. It is currently widely used, in conjunction with other drugs, in the treatment of a wide variety of tumors (such as 2-5).

$$\begin{array}{c} Cl \diagdown \diagup Cl \\ Pt \\ H_3N \diagup \diagdown NH_3 \end{array}$$

I, C - DDP

Most of the research involving use of platinum compounds as antitumoral agents has involved either c-DDP itself or structurally close derivatives. Through studies (such as 6-13), it has been observed that there generally exists a structural window for activity such that the most active antitumoral compounds are a. neutral, b. contain two inert (NH_3) and two labile (Cl) ligands with c. the ligands cis to one another.

The use of c-DDP has been complicated because of negative side effects including gastrointestinal, hematopoietic, immunosuppressive, auditory and renal disfunction (such as 14-17). One method of overcoming some of these negative side effects is to limit the movement of c-DDP through inclusion of it into a polymer thus prohibiting filtration of the polymer by the kidneys, etc.

Two general approaches have been taken. First, attachment of the c-DDP derivative onto a preformed polymer. This was done

II.   III.

by Allcock and coworkers (such as 18-20). Poly[bis(methylamino)-phosphazene],II, is a water soluble polymer with coordination sites on both the side group and chain nitrogen atoms (13,21). Compound II reacts with K_2PtCl_4 and 18-crown-6-ether in organic media giving a coordination complex of structure III (13,19). Compound III shows tumor inhibitory activity towards mouse P38: lymphocytic leukemia and in the Ehrlich Ascites tumor regression test (19). Work is continuing in this area.

A second approach has the c-DDP derivative included as part of the polymer. In 1977 we initially synthesized the first poly-(cis-dichlorodiamineplatinum II) compound (22). Here we will briefly describe the synthesis of such compounds and describe results of preliminary biological assays concentrating on tests related to their potential use as antitumoral agents.

$$K_2PtCl_4 + H_2NRNH_2 \rightarrow \underset{NH_2-R-NH_2)_n}{Cl\diagdown Pt \diagup Cl}$$

IV.

This paper presents preliminary biological assay results related to potential antineoplastic activity for select platinum polyamines of Form IV.

Synthesis and Physical Characterization

Following are typical synthetic and structural characterization procedures more fully described in references 22 and 23. Potassium tetrachloroplatinate (6.22×10^{-4} moles in 10 ml H_2O) and potassium iodide (4.97×10^{-3} moles in 10 ml H_2O) were separately dissolved, treated for ten minutes on a boiling water bath, mixed, heated 20 minutes longer and filtered to remove KCl. The resulting potassium tetraiodiplatinate solution was then mixed with an aqueous solution of 1,6-hexanediamine (7.83×10^{-4} moles in 10 ml of water) to immediately give a curdy, yellow solid, poly(cis-diiodo-1,6-hexamethylenediamine platinum (II)), DIHP, in 74% yield based on initial potassium tetrachloroplatinate. Elemental analysis (performed by Galbraith Labs., Knoxville, Tenn.) was in agreement with a structure of form IV; %C-found=12.0, theory=12.8; %N-found=4.9, theory=4.9; %H-found=2.9, theory= 2.9; %Pt-found=31.3, theory=34.5; %Cl-found 0.1, theory= 0.0; %I-found=49.9, theory=44.9. The following bands, with assignments, were present in infrared spectra of the product (obtained utilizing Perkin-Elmer 457 and 735B spectrophotometers; all values given in cm^{-1}) N-H stretch -2920, 2820; NH_2 bend -1560; CH_2 bend - 1455; CH_2 rock - 720; and Pt-N stretch - 545 and 460 cm^{-1} (24-28). The infrared spectra for the products derived from

K_2PtBr_4 and K_2PtCl_4 are identical from 4000 to 350 cm^{-1} (Figure 1). The Pt-Halide stretch for Pt-I and Pt-Br fall below 250 cm^{-1}, below the capability of the utilized spectrophotometer. The Pt-Cl band is present at 315 cm^{-1}. The UV spectra of the compounds are consistent with a cis-dihalodiamino-platinum (II) compound with bands at (obtained using a Cary 14 UV-visible spectrophotometer; wavelengths given in nm for chloro product) 250, 310, 340 and 390 with identification based on references 12 and 13. NMR spectra (obtained on a Varian EM360A spectrometer) in d_6-DMSO (using tetramethylsilane as a reference) showed three peaks; a broad band at 1.3 ppm (from the eight inner protons of the hexamethylene chain), a broad multiplet at 2.6 ppm (from the methylene protons) and a multiplet at 3.7 ppm. Interference by DMSO contamination of the d_6-DMSO and the low solubility of DIHP in DMSO - about 2% -prevented adequate integration of the NMR bands. A weight average molecular weight (obtained using a Brice-Phoenix 2000 light scattering photometer) of 4200 was found. Analysis employing coupled thermogravimetric-mass spectroscopy of the analogous chloropolyamine showed m/e ions at 35,37 and 70-74 amu indicative of the presence of the chloride moiety and a fragmentation pattern consistent with the presence of hexamethylenediamine.

The above was repeated except utilizing a more nearly equal molar amount of the $PtCl_4^{-2}$ and diamine (2.84 mmole of each reactant) and led to the synthesis (73% yield) of high molecular weight product with a weight average molecular weight of 1.5 x 10^6 via light scattering photometry. Thus product chain length can be easily and effectively controlled through control of the ratio of reactants. All of the tested polyamines are polymeric with degrees of polymerization in excess of 100 (23).

Biological Testing

Figure 2 contains structures of platinum polyamines tested along with a sample designation number. Table 1 contains a brief description of cell lines utilized in the present study.

Table 1. Cell lines utilized in the study of the inhibitory nature of platinum polyamines

Abbreviated Designation(s)	Cell Line
L929 and L	Mouse Connective Tissue (Tumor Cells)
WISH	Human Amnion (Transformed Cells)
HeLa	Human Cervical (Cancer Cells)
VERD	African Green Monkey Kidney (Transformed Cells)

In Vitro. Mouse connective tissue (L cells), human cervical carcinoma (HeLa), and human amnion (WISH) cancer cells were used to determine the toxic and nontoxic levels of several of the

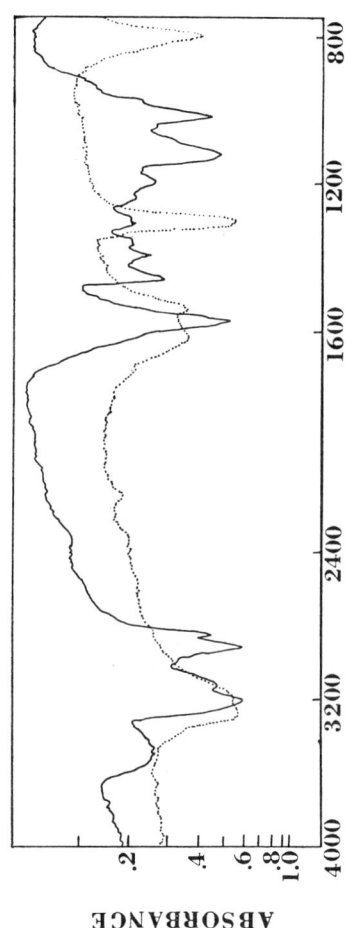

Figure 1. IR spectrum of poly[cis-dichloro(1,6-hexanediamine)platinum(II)] (———) and cis-dichlorodiaminoplatinum(II) (· · ·).

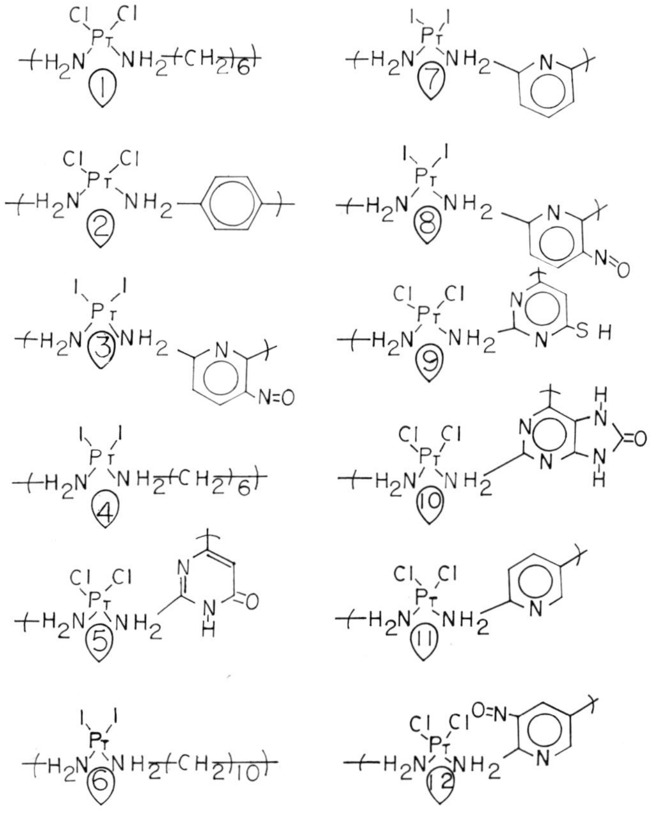

Figure 2. Structures of platinum polyamines utilized in the biological testing.

platinum compounds. Cells were plated into 1 ml wells and, after about 24 hrs, the pure media was removed and 1 ml of media containing the polymer in solution added. The cells were kept under surveillance for a specified time checking for cytopathic effect. The nutrient media contained Dulbeccos modified Eagles medium (DMEM) with 10% calf serum (CS) and 1% penicillin-streptomycin (Pen-Strep). The polymer containing nutrient media contained the afore plus a small amount (less than 1%) of DMSO containing the polyamines. The polyamine remains in solution throughout the testing procedure.

The maximum (threshold) nontoxic concentration level is similar for all compounds tested against tumor and transformed cell lines typically being 1 to 20 µg/ml (Tables 2-4). All of the polyamines exhibit total inhibition (100% cytopathic effect) above 30 µg/ml.

Table 2. Toxicity of Platinum Compounds on Cells in Monolayer Culture

Pt Compound	Concentration (µ g/ml)									
	5		10		20		30		50	
**	HeLa	L929	HeLa	L929	HeLa	L929	HeLa	L929	HeLa	L929
1	0	0	0	0	0	0	25	25	100	*100
2	0	0	0	0	0	0	25	25	100	100
3	0	0	0	0	0	0	25	25	100	100
4	0	0	0	0	0	0	25	25	100	100
5	0	0	0	0	0	0	25	25	100	100

*Cytopathic effect after 24 hr of treatment with the indicated concentration of Pt compound.
**Polymer ID number from Figure 2.

Table 3. Toxicity of Platinum Compounds on Mouse Tumor Cells in Culture.

Polymer Concentration (µg/ml)	Cytopathic Effects After 24hr. Treatment with Polymers						
	6*	7	8	9	10	11	12
60	4	4	4	4	4	4	4
20	4	4	4	4	4	4	4
6.7	3	4	1	+	3	3	3
2.2	1	1	0	0	1	0	0
0.7	0	0	0	0	0	0	0
0	0	0	0	0	0	0	0

Monolayer cultures of L929 (Mouse connective tissue) were treated with the indicated concentrations of each polymer. Toxic effects were recorded at the end of 24 hours.
4 = 100% cell destruction; 3 = 75%, 2 = 50%, 1 = 25%, + = 10%.
*Polymer ID number from Figure 2.

Table 4. Toxicity of Platinum Polymers on Human Amnion (WISH) Cells in Culture.

Polymer Concentration (μg/ml)	Cytopathic Effects After 24 Hours Treatment with Polymer		
	1*	2	4
63	4	4	4
21	4	4	4
7.0	2	4	2
2.3	0	0	0
0.80	0	0	0
0	0	0	0

Monolayer cultures of WISH (human amnion) cells were treated with the indicated concentrations of each polymer. Toxic effects were recorded at the end of 24 hours. 4 = 100% cell destruction, 3 = 75%, 2 = 50%, 1 = 25%, 0 = 0%.
*Polymer ID Number from Figure 2.

Selected polyamines were tested for their ability to alter the normal replication cycle of two representative RNA viruses Encephalomyocarditis (EMC strain MM) and Poliovirus Type I. Monolayers of L929 and HeLa cells were treated with solutions containing specified polyamines. After incubation the cultures were washed and then infected with the RNA virus. After a 24 hours incubation period, the supernatant fluids were collected and assayed for virus (pfu) content.

The data given in Tables 5 and 6 are offered to illustrate the variety of results. Thus for Poliovirus, platinum polyamines can effect an increase or decrease in the growth of the virus or exhibit no effect at concentrations below that where the HeLa cells are affected. Similar results are found for the polyamines in L929 cells infected with EMC strain MM. Thus with treatment for 24 hours, polymers 9,11 and 12 suppressed viral replication (Table 6). With the exception of polymer 6, all of the polymers suppressed EMC viral replication when treatment was extended to 48 hours. Again viral replication was affected at polyamine concentrations well below that necessary to inhibit the L929 cells.

Table 5. Effects of Platinum Compounds on the Viral Replication of Poliovirus in HeLa Cells.

Concentration µg/ml	Compound 1**		2		4	
	Pfu*/ml	%control	Pfu/ml	%control	Pfu/ml	%control
10	3.4×10^6	10	6.5×10^7	200	3.2×10^7	100
20	1.1×10^7	34	3.2×10^7	100	3.2×10^7	100
Control	3.2×10^7	--	3.2×10^7	---	3.2×10^7	---

*plaque forming units
**Polymer identification number from Figure 2.

Table 6. Effect of Platinum Compounds on the Replication of MM Virus in L Cells

Compound	Length of Treatment 2.2µg/ml for (hr)	24hr Virus Yield (pfu/ml)	% Control
6*	24	7.7×10^5	154
	48	4.0×10^5	148
7	24	7.4×10^5	148
	48	1.2×10^5	43
8	24	5.6×10^5	112
	48	2.3×10^5	82
9	24	3.5×10^5	70
	48	1.6×10^5	57
10	24	5.2×10^5	104
	48	1.6×10^5	57
11	24	2.6×10^5	52
	48	1.2×10^5	43
12	24	3.8×10^5	76
	48	2.8×10^5	75
Control	24	5.0×10^5	N/A
	48	2.8×10^5	N/A

*Polymer identification number from Figure 2.

Monolayer cultures of L929 cells were treated for 24 and 48 hours with 2.2 µg/ml of each polyamine. The cultures were then washed and infected with MM virus. After a 24 hour incubation period, the supernatant fluids were collected and assayed for virus pfa content.

Thus preliminary studies show that while there was no overt evidence of activity by the platinum polyamines on the cell cultures (CPE) at the employed polymer concentrations, there was biological activity in that the polymers were inhibiting or stimulating viral replication.

The RNA viruses EMC-MM and Poliovirus Type 1 are picoviruses with biological responses similar to other RNA viruses. Biological repsonses to agents by these two viruses is applicable to other RNA viruses. Implications of control of viral activity at low concentrations of drug is applicable to not only (potential) control of tumor growth, but also to control of RNA virus associated diseases including Polio, Encephalipis, the majority of "common colds," and most gastro intestinal viral infections. Thus drug control of RNA viruses is significant with broad implications.

In vivo. Preliminary experiments with mice show that they can tolerate a dosage of 400 µg of compound #4 (highest concentration tested) with no apparent ill effects. This dose is in excess of tenfold greater than that necessary to destroy either HeLa or L929 tumor cells. Further 500 µg of polymer 1 was given intraperitoneally at 3 day intervals to mice for a 15 day period with no apparent ill effects.

Thus mice are able to tolerate relatively high dosages of the platinum polyamines without overt ill effects.

Literature Cited

1. B. Rosenberg, L. VanCamp and T. Krigas, Nature, 205, 698 (1965).
2. D. Higby, H. Wallace, D. Albert and J. Holland, J. of Urology, 112, 100 (1974).
3. E. Wiltshaw and T. Kroner, Cancer Treatment Reports, 60, 55 (1976).
4. J. Hill, E. Loeb, A. Pardue, A. Khan, N. Hill, J. King and R. Hill, J. Clinical Hematology and Oncology, 7, 681 (1977).
5. A. Yagoda, R. Watson, H. Grabstald, and W. Whitmore, Proceedings of the American Assoc. Cancer Res., 17, 296 (1976).
6. T. Conners, M. Jones, W. Ross, P. Braddock, A. Khokhar and M. Tobe, Chemico-Biological Internations, 5, 415 (1972).
7. M. Cleare and J. Hoeschele, Platinum Metals Rev., 17, 2 (1973).

8. P. Schwartz, S. Meischen, G. Gale, L. Atkins, A. Smith and E. Walker, Cancer Treatment Repts., 61, 1519 (1977).
9. J. Hugheey, "Inorganic Chemistry: Principles of Structure and reactivity," Harper and Row, N.Y., pages 423-425, 1972.
10. F. Cotton and G. Wilkinson, Advanced Inorganic Chemistry, Third Ed., Interscience Pubs., N.Y., pages 665-669, 1972.
11. G. Kauffman, Inorganic Synthesis, 1, 249 (1963).
12. H. Ito, J. Fugita and K. Sato, Bull. Chem. Soc. Japan, 40, 2584 (1967).
13. J. Chatt, G. Gamlen and L. Orgel, J. Chem. Soc., 486 (1958).
14. J. Ward and K. Fauvie, Toxicology and Applied Pharmacology, 38, 535 (1976).
15. J. Gottlieb and B. Drewinko, Cancer Chemotherapy Repts., 59, 621 (1975).
16. I. Krakoff and A. Lippman, Recent Results in Cancer Research, 48, 183 (1974).
17. A. Khan, J. Hill, W. Grater, E. Loeb, A. MacLellan and N. Hill, Cancer Research, 35, 2766 (1975).
18. A. Allcock, Science, 193, 1214 (1976).
19. H. Allcock, "Organometallic Polymers," (Editors C. Carraher, J. Sheats and C. Pittman), Chpt. 28, Academic Press, N.Y., 1978.
20. H. Allcock, R. Allen and J. O'Brien, J. Amer. Chem. Soc., 97, 39: 4 (1977).
21. H. Allcock, W. Cook and D. Mack, Inorg. Chem., 4, 2584 (1972).
22. C. Carraher, J. Schroeder, D. Giron and W. Scott, J. Macromol. Sci.-Chem., A15(4), 625 (1981)..
23. C. Carraher and W. Scott, J. Macromol. Sci.-Chem., in preparation.
24. G. Barrow, R. Krueger and F. Basolo, J. Inorg. and Nucl. Chem., 2, 340 (1956).
25. R. Berg and K. Rasmussen, Spectrochimica Acta, 29A, 319 (1973).
26. L. Segal and F. Eggerton, Applied Spectroscopy, 15, 116 (1961).
27. G. Watt, B. Hutchinson and D. Klett, J. Am. Chem. Soc., 89, 2007 (1967).
28. Y. Kharitonov, I. Dymina and T. Leonovan, Russian J. of Inorganic Chemistry, 13, 709 (1968).

RECEIVED July 7, 1981.

18

Controlled Release of Anticancer Agents That Are Complexes of *cis*-Diamminedichloroplatinum(II) and α-Hydroxyquinones

SEYMOUR YOLLES, ROSETTE M. ROAT, MARIO F. SARTORI, and CATHARINE L. WASHBURNE

University of Delaware, Department of Chemistry, Newark, DE 19711

> Complexes between the square planar anticancer agent cis-diamminedichloroplatinum (II), (cis-Pta_2Cl_2), and α-hydroxyquinones are reported. A complex between cis-Pta_2Cl_2 and the anthracycline antibiotic, doxorubicin, contains five cis-Pta_2Cl_2 units per doxorubicin molecule. Ratios of cis-Pta_2Cl_2 to α-hydroxyquinones range from 3/1 for cis-Pta_2Cl_2/juglone complexes to 6/1 for cis-Pta_2Cl_2/quinizarin complexes. Ultraviolet and infrared data are reported for the complexes as well as thin layer chromatography, conductivity, and molecular weight results. The results indicate chelation of platinum to oxygen atoms of the α-hydroxyquinones with concomitant stacking of cis-Pta_2Cl_2 units. The reaction leading to dimerization has the potential for further chain extension although no such compounds have as yet been isolated. In vivo experiments with cis-Pta_2Cl_2/doxorubicin and with cis-Pta_2Cl_2/quinizarin show the efficacy of these complexes as a new time-release system for delivering anticancer agents.

Several papers (*1*, *2*, *3*) from this laboratory have dealt with the development of a system for delivering anticancer agents at a constant rate over a prolonged period, perhaps as long as several months. This system consists of incorporating the drug in a polymeric matrix, shaping the composite into convenient form (such as beads or powder) and implanting the composite into the body tissue of animals. The drugs diffuse continuously from the interior of the polymer to the outer surface, where they are mechanically swept away by body fluid. The mechanism of migration of the drug is that of diffusion and the thermodynamic driving force is the concentration gradient (*3*).

0097-6156/82/0186-0233$5.00/0
© 1982 American Chemical Society

Previously reported (2) in vitro and in vivo experiments with composites of poly(lactic acid) (PLA) with cyclophosphamide, with cis-diamminedichloroplatinum (II) (cis-Pta_2Cl_2) (4), and with cis-Pta_2Cl_2-doxorubicin (DOXO) (5) have demonstrated the feasibility of this system, especially their drug release ability and the reduced toxicity of the platinum compound.

During the preparation of the composite containing PLA, cis-Pta_2Cl_2 and DOXO, a distinctive color change to intense purple was observed (2). In view of the enhanced anticancer activity shown by this composite in animals over those containing the separate drugs, it was decided to isolate and characterize the purple material, believed to be a chemical complex of the platinum with DOXO and, in addition, to investigate the efficacy of this complex in in vivo experiments.

To elucidate the binding site, several α-hydroxyquinone analogues of DOXO were complexed with cis-Pta_2Cl_2.

This paper reports the syntheses and characterization of complexes of cis-Pta_2Cl_2 with 5-hydroxy-1,4-naphthoquinone (HNQ), 5,8-dihydroxy-1,4-naphthoquinone (DHNQ), 1,4-dihydroxy-9,10-anthraquinone (DHAQ) and with DOXO. In addition it reports preliminary results of in vivo experiments performed with complexes cis-Pta_2Cl_2/DOXO and cis-Pta_2Cl_2/quinizarin.

Several metal complexes (Be, Co, Cu, Fe, Mg, Ni) with DOXO (6, 7) and with hydroxy-naphthoquinones and -anthraquinones (8, 9, 10) have been recently reported. To our knowledge no study of platinum complexes with α-hydroxyquinones has been published to date, with the exception of an unsuccessful attempt to prepare a Pt/5,8-dihydroxy-1,4-naphthoquinone (8).

Experimental Section

General Information. All reactants were dried at room temperature under vacuum in an Abderhalden apparatus to constant weight. All experiments were carried out in a dry box; continuously under nitrogen. Elemental analyses were performed by Microanalysis, Wilmington, DE, all compounds prepared yielded satisfactory analyses (see Table I). The UV spectra were obtained on samples dissolved in DMF with a NSI Hitachi computerized #100-80 Ultraviolet Spectrophotometer. The IR spectra were recorded on a Perkin Elmer 180 Recording IR Spectrophotometer in CsI pellets (except for DOXO and DOXO complex where KBr discs were used). Conductivity measurements were taken on a Beckman Model RC16B2 conductivity bridge using a Fisher conductivity cell (cell constant = 0.1799, determined on standard KCl solutions). ^1H NMR spectra were obtained in d_7DMF at 90MHz on a Bruker HFX-10 spectrometer. Thin layer chromatographs on silica gel 60 supports from 0.1 to 1.0 mM DMF solutions of starting materials and products were developed in 9/1 acetone/HCl, and visualized with 2% $SnCl_2$ in 1M HCl spray.

Experiments In Vivo. The standard assay of the Ascites Sarcoma in the ICR mice was used (2). There were six mice in each cage, with two cages for the negative controls, two cages for the positive controls and one cage for each dose level of drug tested. Tumor cell injections were performed on day 0 and drug injection on day 1, both given intraperitoneally. Suspensions in carboxymethyl cellulose (CMC-7LF, Hercules, Inc., Wilmington, DE) of complexes, as listed in Table IV, with saline as a carrier were used.

Complex I: cis-Pta_2Cl_2/5-hydroxy-1,4-napthoquinone. In a typical preparation, cis-Pta_2Cl_2 (0.50 g, 1.66 mmol) was dissolved in 30 ml N,N-dimethylformamide (DMF) and added to HNQ (0.14 g, 0.80 mmol) dissolved in 10 ml DMF. The reactants were combined in a 120 ml brown glass jar which contained 83.5 gms of 5 mm glass beads. The jar was sealed under N_2, then placed on a ball mill (1750 rpm) to rotate for 120 hrs. The ball milled suspension (brown-purple in color) was centrifuged to remove a small amount of solids. The main product was precipitated by addition of 80 ml methylene chloride ($MeCl_2$) to yield a brown-purple solid collected on a frit-glass filter. Crude product was recrystallized from 10-15 ml DMF with precipitation by excess $MeCl_2$. Yield: 0.59 gm (90%) of brown-purple product.

Table I Elemental Analyses

	C	H	Cl	N	Pt	
Complex I	11.77	2.27	15.63	8.24	57.37	a
cis-Pta_2Cl_2/HNQ	11.68	2.69	15.71	8.61	55.51	b
$C_{10}H_{22}Cl_{4.5}N_6O_3Pt_3$						
Complex II	9.22	2.16	14.97	8.60	59.94	a
cis-Pta_2Cl_2/DHNQ	8.68	2.40	14.75	7.51	58.68	b
$C_{10}H_{30}Cl_{5.5}N_8O_4Pt_4$						
Complex III	8.38	1.90	21.20	6.98	58.33	a
cis-Pta_2Cl_2/DHAQ	8.68	2.48	20.10	6.52	56.76	b
$C_{14}H_{36}Cl_{12}N_{10}O_4Pt_6$						
Complex IV	16.56	3.32	14.39	7.85	49.50	a
cis-Pta_2Cl_2/DOXO	15.86	3.21	14.15	7.52	47.64	b
$C_{27}H_{56}Cl_8N_{11}O_{11}Pt_5$						

a Calculated
b Found

Complex II: cis-Pta_2Cl_2/5,8-dihydroxy-1,4-naphthoquinone. Following the procedure described above, 5,8-dihydroxy-1,4-naphthoquinone (0.039 g, 0.20 mmol) and cis-Pta_2Cl_2 (0.373 g, 1.24 mmol) were condensed to give 0.40 g (yield 97%) of compound II as navy blue product.

Complex III: cis-Pta$_2$Cl$_2$/1,4-dihydroxy-9,10-anthraquinone. Following the procedure described above, 1,4-dihydroxy-9,10-anthraquinone (0.062 g, 0.26 mmol) was reacted with cis-Pta$_2$Cl$_2$ (0.50 g, 1.66 mmol) to give 0.29 g (yield 52%) III as purple product.

Doxorubicin Base from Doxorubicin Hydrochloride. Doxorubicin hydrochloride was converted to the free base by bubbling gaseous ammonia through a capillary pipet into a stirred suspension of doxorubicin hydrochloride (0.050 g, 0.08 mmol) in 15 ml of an 8:1 mixture of methylene chloride and methanol at 0°C. The bubbling of ammonia and the stirring were continued until the mixture was red-orange. After the ammonium chloride was removed by centrifugation, excess solvent and ammonia were removed on a Rotovap. DMF was added to give a solution of 0.047 g (0.086 mmole) of doxorubicin base in 100 ml of solvent.

Complex IV: cis-Pta$_2$Cl$_2$/Doxorubicin. Following the procedure described above for complex I, cis-Pta$_2$Cl$_2$ (0.104 g, 0.34 mmol) was condensed with doxorubicin (0.047 g, 0.086 mmol) to give 0.103 g (68.7% yield) of compound IV as purple product.

Results and Discussion

Complexes of cis-Pta$_2$Cl$_2$ with HNQ, DHNQ, DHAQ and DOXO have been obtained. All these complexes are similar in physical form: fine powder (solids) of intense purple to blue color; poorly soluble in common organic solvents, giving purple or blue solutions; none sublime on heating; all decompose before melting. U.V. and I.R. spectra of the parent ligands and platinum complexes are reported in Tables II and III.

α-Hydroxyquinones exhibit a so-called chelated H absorbance in the UV, caused by H bonding of the α-hydroxyl H to the keto oxygen of the quinone (11). In our complexes with platinum, this absorbance disappears, with concurrent appearance of a new absorbance at longer wavelengths characteristic of the α-hydroxyquinone dianion (11).

Retention of the 302 nm band supports the presence of cis-Pta2Cl2 units in the structures of these complexes. It is assumed that these units are stacked in a way similar to that of the Magnus' green salt (12). Hydrolysis of a cis-Pta$_2$Cl$_2$/DHAQ complex in dilute HCl, followed by thin layer chromotography has identified 75% of the Pt containing fraction as pure cis-Pta$_2$Cl$_2$, the other 25% being the fractions which have lost chloride ligand. Pure quinizarin was also identified as a component of the hydrolyzate by thin layer chromotography.

The IR spectra of the quinone complexes show a shift of the $\nu(C=O)$ band stretching frequency to lower wavenumbers in comparison with the corresponding parent hydroxyquinones. This shift indicates coordination or chelation of Pt(II) to oxygen of the hydroxyquinone as shown by Pierpont (10) for the copper (II) complexes of 1,5-dihydroxy-1,4-naphthoquinone.

Table II Ultraviolet Spectra of Hydroxyquinone Ligands and Their Corresponding Pt(II) Complexes[1]

Compounds	Absorption (log ε) nm
HNQ	425(3.6), 260(4.2)
HNQ Complex	485, 380, 260(4.2)
DHNQ	515(4.2), 482(4.2), 270(4.4)
DHNQ Complex	620, 570, 302, 271(4.4)
DHAQ	478(3.8), 278(4.0)
DHAQ Complex	580, 540, 302, 278
DOXO	530(2.7), 495(2.9), 475(2.9), 265(3.0)
DOXO Complex	592(2.7), 535(2.8), 506(2.8), 265(3.3)

[1]Absorption maxima cis-Pta_2Cl_2 370,302 nm.

Table III IR Absorption (cm^{-1}) of hydroxyquinone ligands and their Pt(II) complexes

Cmpds	Chelate OH	Chelate OX	ν(C=O)	C-O-Pt	Pt-O
DHNQ	3010[a]	--	1600	--	--
DHNQ Cmplx	--	2910[b]	1560	1030-1170	445
DHAQ	3000[a]	--	1626,1590	--	--
DHAQ Cmplx	--	2910[b]	1605,1575	1000-1150	445
DOXO	2910	--	1620-1635	--	--
DOXO Cmplx	--	--	1620-1630	1050-1150	450

[a]Hadzi, D.; Sheppard, N., Trans. Far. Soc., 1954, 50, 911
[b]reference 13

Broad, strong absorption bands at 1000-1500 cm^{-1} reported (13) to be indicative of ether type linkages, are tentatively assumed to be due to C-O-Pt linkages. Absorbance at 445-450 cm^{-1} is analogous to that reported by Nakomoto for Pt-O bonds in Pt(II) acetylacetonates (14). All complexes also exhibit IR (in cm^{-1}) absorbances reported for cis-Pta_2Cl_2 by Poulet (15): ν (NH_3): 3280, 3200; δ (NH_3): 1620, 1290; ν (Pt-Cl); 315; ν (Pt-N): 240.
^1H NMR spectra of DHAQ (δ = 12.8, chelated OH; δ = 8.3-8.4, aromatic H; δ = 7.5, 2,3H) and cis-Pta_2Cl_2 (δ = 4.2, N-H) were straightforward, however, NMR of the complex showed retention of

the N-H peak of cis-Pta$_2$Cl$_2$ only, plus several unidentified upfield resonances. Experiments are currently underway to show that these unidentified peaks are isotropically shifted DHAQ resonanc s caused by paramagnetism in the complex.

Attempts at obtaining single crystals for X-ray structure determination have failed thus far. Low solubility of the complexes in solvents suitable for vapor pressure osmometry have made molecular weight determination difficult. One result for complex III, cis-Pta$_2$Cl$_2$/DHAQ, in acetone yielded a M.W. of 2270 gm/mole, vs a calculated M.W. from elemental analyses of 2006 gm/mole.

Based on data given, a probable structure of Complex IV, cis-Pta$_2$Cl$_2$/DOXO, is shown in Figure 1. Conductivity of the cis-Pta$_2$Cl$_2$/DHAQ complex increased from zero, immediately after solution in DMF, to values which varied with concentration in the manner of a 2:1 electrolyte (as compared to K$_2$PtCl$_6$ by the method of Feltham and Hayter (16). Probable structure of the electrolyte differs from that in Figure 1 in that the two five-co-ordinate platinum centers become four-co-ordinate through loss of chloride anions, resulting in a dipositive complex ion containing platinum co-ordinated to the hydroxyquinone portion of DOXO. Fractional chlorine content from elemental analyses calculations for HNQ and DHNQ complexes indicates that either one or two chloride ligands are lost from cis-Pta$_2$Cl$_2$ on complexation.

In Vivo Experimental Results

Preliminary in vivo results (see Table IV) show the feasibility of this potential time-release system for delivering anticancer agents. cis-Pta$_2$Cl$_2$/DOXO complex was significantly better than the positive control (7 mg cis-Pta$_2$Cl$_2$/kg mouse). The improvement in life span compared with the positive control is substantial and the toxic effect of the platinum compound has certainly been decreased. These tests show that it is possible to use a dose of complex three times the toxic level of cis-Pta$_2$Cl$_2$, because of slow release from cis-Pta$_2$Cl$_2$/DOXO complex.

Preliminary in vivo results (see Table IV) show that the complex formed between cis-Pta$_2$Cl$_2$ and quinizarin was more efficacious than the positive control. It did not have quite the degree of increased life span shown by the cis-Pta$_2$Cl$_2$/DOXO complex. On the other hand, as before with the DOXO complex, fairly high levels of cis-platinum upwards of 20 mg/Kg mouse could be used as compared to a toxic level of 7 mg/Kg mouse for cis-Pta$_2$Cl$_2$ alone.

This paper has reported only products of the reaction which were isolated and characterized. However, the potential for further chelation and therefore doubling or trebling the molecular weight to polymer proportions exists. This suggests the possibility of a polymeric drug moiety which would be inherently slow releasing in vivo. It is planned to continue this line of research in the future.

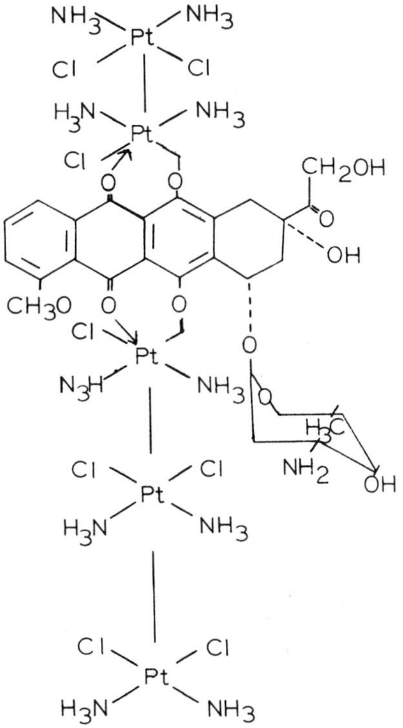

Figure 1.   cis-Diamminedichloroplatinum(II)/DOXO complex.

Table IV

In Vivo Test Results of Complexes

Compound	Dose mg Compound/ kg Mouse	Avg. Day of Death	% Increased Life Span	Number of Cures[2] (out of 6)
Negative Control[1]	0	17.3	--	-
Positive control cis-Pta$_2$Cl$_2$	7	31.3	80	4
cis-Pta$_2$Cl$_2$ + doxorubicin	5	34.2	97	4
	10	36.0	108	6
	15	34.8	101	1
	20	26.2	51	1
Negative control	0	13.8	--	-
Positive control cis-Pta$_2$Cl$_2$	7	25.0	81	1
cis-Pta$_2$Cl$_2$ + quinizarin	8	28.0	107	1
	12.5	26.0	88	0
	25	16.0	16	0
	50	2.75	-80	
	100	3.8	-72	

[1] Negative control: Mice injected only with cancer cells.

[2] Cure: The National Cancer Institute uses the term "cure" for those animals which live beyond 60 days.

Acknowledgments

We acknowledge generous gifts of doxorubicin (Adriamycin) by Adria Laboratories, Inc.; cis-Pta_2Cl_2 by Johnson Matthey, Inc. Thanks to H. Blount, E. Bancroft, G. McIntire for electrochemical analyses, J. Wells and J. Murtha for compound syntheses and TLC.

Literature Cited

1. Yolles, S.; Leafe, T.D.; Meyer, F.J. J. Pharm. Sci. 1975, 64, 115.
2. Yolles, S.; Morton, J.; Rosenberg, B. Acta Pharm. Suecica 1978, 15, 382.
3. Yolles, S.; Sartori, M.F. "Degradable Polymers for Sustained Drug Release" in Drug Delivery Systems, R.L. Juliano, Oxford University Press, Inc., New York, 1980; p. 84.
4. This drug has recently been approved by the FDA.
5. An anthracycline antibiotic isolated from Streptomyces peucetius.
6. Gosalvez, M.; Blanco, M.F.; Vivero, C.; Valles, F., Europ. J. Cancer 1978, 14, 1185.
7. Greenaway, F.T.; Dabrowiak, J.C., ACS Inorganic Abstract #169 (Sept. 1979).
8. Bottei, R.S.; Gerace, P.L., J. Inorg. Nucl. Chem. 1961, 23, 245.
9. Coble, H.D.; Holtzclaw, H.F., Jr., J. Inorg. Nucl Chem. 1974, 36, 1049.
10. Pierpont, C.G.; Francesconi, L.C.; Hendrickson, D.N., Inorg. Chem. 1978, 17, 3470.
11. Thomson, R.H. "Naturally Occurring Quinones" (2nd Ed.) Academic Press, New York, 1971; pp. 39-92.
12. Cotton, F.A.; Wilkinson, G. "Advanced Inorganic Chemistry" (3rd Ed.) Wiley and Sons, New York, 1972; pp. 1034, 1035, 1036.
13. Bellamy, L.J. "Infrared Spectroscopy of Complex Molecules" (2nd ed. reprint) Wiley and Sons, New York, 1959.
14. Nakomoto, K.; McCarthy, P.J. "Spectroscopy and Structure of Metal Chelate Compounds" Wiley and Sons, New York, 1968; pp. 251 ff.
15. Poulet, H.; Delorme, P.; Mathieu, J.P. Spectrochim. Acta 1964, 20, 1855.
16. Feltham, R.D.; Hayter, R.G. J. Chem. Soc., 1964, 4587.

RECEIVED October 1, 1981.

^{13}C NMR Studies of the Structure of the Divinyl Ether–Maleic Anhydride Cyclic Alternating Copolymer: A Biologically Active Agent

WALTER J. FREEMAN and DAVID S. BRESLOW
Hercules Incorporated, Research Center, Wilmington, DE 19899

The copolymer has been shown to contain tetrahydrofuran and tetrahydropyran rings in a ratio of 0.8 to 1. Resonances have been assigned by the use of single frequency off resonance decoupling (SFORD), suppressed nuclear Overhauser effect (NOE) combined with a relaxation reagent, and comparison with model compounds. Published spectra, which have been interpreted as demonstrating the presence of only 5-membered rings, also show the unique 6-membered ring methylene group. Available evidence is insufficient to assign stereochemistry.

Maleic anhydride and divinyl ether form a cyclic alternating copolymer in a 2 to 1 ratio[1]. This polymer has a broad spectrum of biological activities[2], and is at present undergoing preliminary clinical evaluation as an antitumor agent. The structure of the 2 to 1 copolymer has proven to be somewhat elusive. Various techniques have been used in the past to study and characterize this copolymer. Recently, two studies have appeared which focus attention on the ^{13}C and proton NMR properties of the copolymer. Kunitake and Tsukino[3] have presented evidence based on the ^{13}C NMR data that the copolymer may consist primarily of 5-membered or 6-membered bicyclic structures, depending on the method of polymerization; the 6-membered bicyclic structure has been promoted by Butler[4]. The symmetrical or 5-membered ring structure was concluded by Kunitake to be formed when the polymerization was done in chloroform. Moreover, chain methylenes were calculated to be in the cis-configuration on the tetrahydrofuran rings and trans on succinic anhydride units along the chain backbone. When the copolymer was prepared in a mixed-solvent system consisting of acetone and carbon disulfide, the tetrahydropyran bicyclic unit was found to be favored.
 Butler and Chu[4] have studied the copolymer prepared in cyclohexanone by means of high (300MHz) and low (60MHz) field

0097-6156/82/0186-0243$5.00/0
© 1982 American Chemical Society

proton NMR spectroscopy. These workers have concluded that the pyran or unsymmetrical structure is exclusively formed. Further, the pyran anhydride ring junction was concluded to possess a trans-geometry. The proton NMR study was supported by comparison of copolymers prepared in turn with bis(2,2-dideuterovinyl) ether and maleic anhydride, and divinyl ether and 3,4-dideuteromaleic anhydride. These workers further found that the ^{13}C spectra of the copolymers prepared in cyclohexanone were invariant with temperature. This observation was considered by them as additional evidence supporting the energetically favored 6-membered structure. Based on these data, a mechanism using HOMO-LUMO arguments was put forth to rationalize the exclusive formation of the tetrahydropyran structure.

Results

In this paper we present evidence, based primarily on ^{13}C NMR data that, for a copolymer prepared in toluene or benzene, a mixture of the symmetrical and unsymmetrical bicyclic structures is obtained. Our studies rest largely on heretofore unavailable model compounds which are representative of the substructural units.

Although NMR spectra can be taken of the anhydride in acetone, much better spectra can be obtained if the copolymer is dissolved in water (or for the purposes of the NMR experiment in D_2O), as shown in Figure 1. Using the labeling scheme of Kunitake we see, from high field to low, peak A (31.3 ppm), B (35.3 ppm), and C (44.7 ppm); a group of six peaks D-E (50.5-53.1 ppm); a group of four peaks, F (76.6-79.8 ppm); and two types of carbonyl, G (173.9-174.3) and H (177.2 ppm). The two small peaks at 25.3 and 68.5 ppm are due to residual tetrahydrofuran. Note that peak B at 35.3 pm is rather distinct and well resolved from peak A at 31.3 ppm. Dioxane (not shown in Figure 1) was used as an internal chemical shift reference (67.0 ppm).

We have assigned all of the ^{13}C peaks which appear in the spectrum of the copolymer and are in general agreement with the assignments of Kunitake (see Table I). However, based on model compound data (vide infra), we find that peak B is due to a 6-membered tetrahydropyran ring methylene carbon. On the basis of this evidence, as well as additional evidence to be presented, we conclude that the copolymer as prepared in our hands consists of a mixture of 5- and 6-membered rings. We note that the appearance of the same peak, i.e., B in the published spectra of Kunitake, demands that he also found a mixture of structures in the same copolymeric material. From our ^{13}C NMR data, the divinyl ether-maleic anhydride copolymer consists of both 5- and 6-membered cyclic structures in the ratio of about 0.8 to 1.

Assigning the ^{13}C NMR peaks of the copolymer has not been a trivial problem by any means. The assignments rest on a variety of bases. These include ^{13}C shifts for model compounds, SFORD

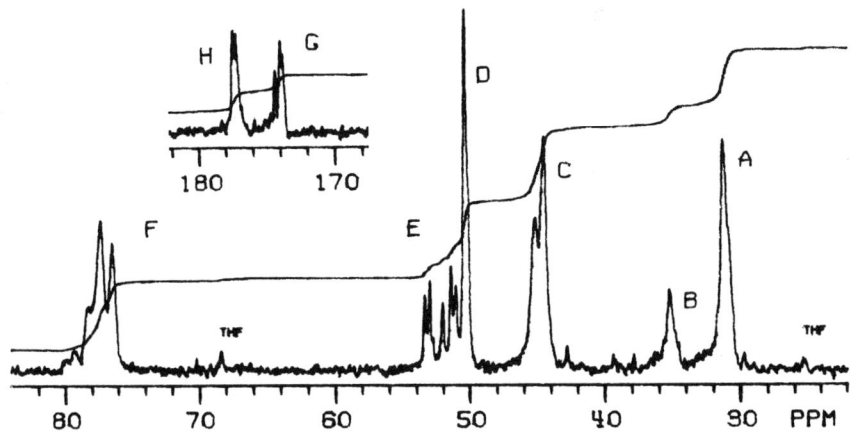

Figure 1. The 90-MHz ^{13}C NMR spectrum of hydrolyzed divinyl ether–maleic anhydride copolymer in D_2O at 58°C taken on a Nicolet NT-360WB.

TABLE I

SUMMARY OF ^{13}C NMR DATA FOR HYDROLYZED DIVINYL ETHER-MALEIC ANHYDRIDE COPOLYMER

PEAK	A	B	C	D AND E	F	G	H
SHIFT (PPM)	31.3	35.5	44.7	50.5-53.1	76.6-79.8	173.9-174.3	177.2
MULTIPLICITY*	T	T	D	D	D	S	S
NO. OF CARBONS							
CALC.	3	1	5	3	4	4	4
FOUND**	2.6	1	4.6	2.6	3.6	3.6	3.6

(SPECTRA RUN AS D_2O SOLUTIONS; INTERNAL REFERENCE DIOXANE ASSIGNED TO 67.0 PPM.)

* T - TRIPLETT (CH_2) D = DOUBLET (CH) S = SINGLET (C)

** FROM NOE SUPRESSED EXPERIMENTS FOR RATIO OF STRUCTURES SHOWN USING AREA OF PEAK AT 35 PPM AS ONE CARBON.

spectra, and a quantitative determination of the relative number of carbons present. The latter was done by means of a combination of a suppressed NOE and a shift or relaxation agent experiment.

Table I shows by means of a SFORD experiment that the peaks at 31.3 and 35.2 ppm appear as triplets and are due to methylene carbons. The rest of the aliphatic carbons are all doublets or methine carbons. We find also from suppressed NOE experiments (Table I) that the peak areas are as follows: the methylene resonance at 31.3 ppm represents 3 carbons; the methylene resonance at 35.2 ppm corresponds to one carbon; the methine resonance at 44.7 ppm represents 5 carbons; the peak appearing at about 50.5 ppm accounts for 3 carbons; and the methine resonance at about 77 ppm represents 4 carbons. The number of carbons present can be assigned either by assuming that the resonance at 77.4 ppm represents 4 carbons or that the resonance at 35.3 ppm represents one carbon.

Discussion

In order to assign the ^{13}C resonances unequivocally, two model compounds were prepared, a 2,6-dimethyltetrahydropyran-3,4-dicarboxylic acid and a 2,5-dimethyltetrahydrofuran-3,4-dicarboxylic acid; the syntheses and ^{13}C structural assignments are shown in Figure 2. One distinct stereoisomer of the tetrahydropyran derivative was produced as a result of the synthesis. The relative stereochemistry of the four functional groups of the tetrahydropyran has been established by means of a proton NMR study. On the basis of the NMR coupling constants and the Karplus relationship, we have determined that this tetra substituted tetrahydropyran is in the all trans-configuration.

The second model compound, 2,5-dimethyltetrahydrofuran-3,4-dicarboxylic acid, was found to consist of three isomers (plus a partially hydrogenated material). We were unable to assign the stereochemistry of this model unequivocally. We also used as a model 2,3-dimethylsuccinic acid which was a mixture of the racemic and meso forms. On this basis, one can show that the resonances of the carbons alpha to the anhydride or acid group along the backbone occur at 44.6 ppm and are approximately equal to that found in the copolymer.

Upon close inspection of the assignments for the copolymer (shown in Figure 1), one can see that the carbon resonances of the 5- and 6-membered rings are overlapped. The only unique carbon resonance is that of the 6-membered ring, and corresponds to peak B in the spectrum of the copolymer. Thus, with the aid of the model compounds and the relative areas found in the copolymer, it is possible to show that the copolymer must consist of a mixture of 5- and 6-membered rings in an approximate 0.8 to 1 ratio.

We feel that the fine structure associated with some of the peaks in the ^{13}C spectrum of the copolymer is undoubtedly due

A

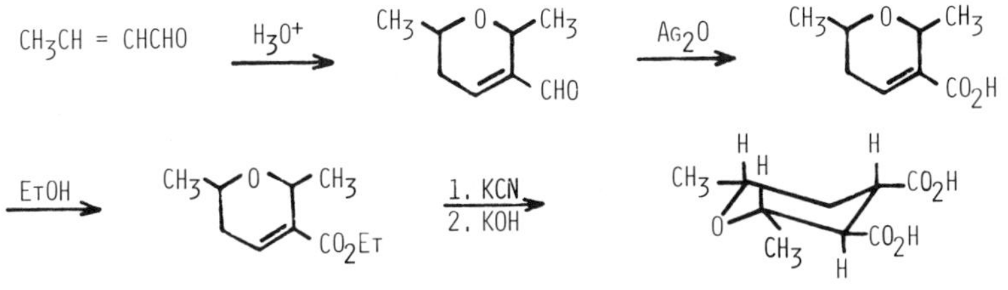

[Scheme A: CH₃CH=CHCHO →(H₃O⁺) 2,6-dimethyl-dihydropyran-3-carbaldehyde →(Ag₂O) 2,6-dimethyl-dihydropyran-3-carboxylic acid; →(EtOH) ethyl ester →(1. KCN, 2. KOH) 2,6-dimethyltetrahydropyran-3,4-dicarboxylic acid]

^{13}C CHEM SHIFTS (22.6 MHz): IN D$_2$O 19.8 (CH$_3$ ON C-2), 21.2 (CH$_3$ ON C-6), 34.6 (C-5), 44.3 (C-4), 51.8 (C-3), 74.1 (C-6), 74.9 (C-1), 177.3 (C=O ON C-3), AND 177.8 PPM (C=O ON C-4). ^{1}H CHEM SHIFTS OF RING PROTON (90 MHz) FOR DIETHYL ESTER IN C$_6$D$_6$: 1.42 (H-5A, J$_{4,5A}$ = 12.5 Hz, J$_{5E,5A}$ = -12.5 Hz, J$_{5A,6}$ = 11.5 Hz), 2.08 (H-5E, J$_{4,5E}$ = 3.8 Hz, J$_{5E,6}$ = 2.0 Hz), 2.36 (H-3, J$_{2,3}$ = 9.5 Hz, J$_{3,4}$ = 11.3 Hz), 2.99 (H-4), 3.62 (H-6) AND 3.64 PPM (H-2).

B

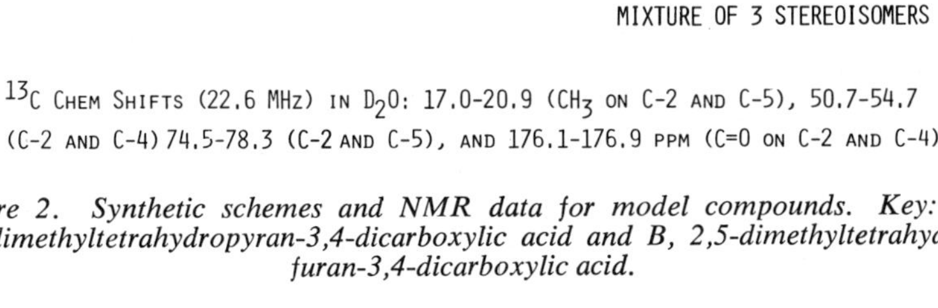

[Scheme B: CH$_3$COCHCO$_2$Et ⁻Na⁺ →(I$_2$) (CH$_3$COCHCO$_2$Et)$_2$ →(P$_2$O$_5$) 2,5-dimethyl-3,4-bis(ethoxycarbonyl)furan →(1. H$_2$, 2. NaOH) 2,5-dimethyltetrahydrofuran-3,4-dicarboxylic acid]

MIXTURE OF 3 STEREOISOMERS

^{13}C CHEM SHIFTS (22.6 MHz) IN D$_2$O: 17.0-20.9 (CH$_3$ ON C-2 AND C-5), 50.7-54.7 (C-2 AND C-4) 74.5-78.3 (C-2 AND C-5), AND 176.1-176.9 PPM (C=O ON C-2 AND C-4)

Figure 2. *Synthetic schemes and NMR data for model compounds. Key: A, 2,6-dimethyltetrahydropyran-3,4-dicarboxylic acid and B, 2,5-dimethyltetrahydrofuran-3,4-dicarboxylic acid.*

to stereochemical differences at ring junctures. Both Butler and Kunitake make some stereochemical assignments based on what we feel is rather tentative data. On the basis of our model compound data, we would agree that stereochemical implications are contained within the ^{13}C data, but we do not feel that anyone at this time has a firm enough spectroscopic basis to make absolute pronouncements about stereochemistry.

Relevance of Models. In our studies, we measured the spectra of model compounds as opposed to the approach of Kunitake, in which he calculated chemical shifts utilizing more generalized methods. We do not feel that a calculation allows one to make a fine enough distinction, particularly when the models used as a basis for the calculation are well removed from those needed for the actual structure study. As an example, Kunitake argues that for a succinic anhydride substructure, carbons 2 and 3 of the succinate ring would be expected to resonate at 38 ppm. The method of calculation used to arrive at this shift is shown in Figure 3. The 38 ppm value represents an upfield shift of -5.1 ppm to form the cis compound. Using the same reasoning the anhydride carbonyls and the 2,3-cis-diisopropyl anhydride are found at 169.1 ppm for an upfield shift of -2.1 ppm for the carbonyls. The calculation for the cis structure is discounted by Kunitake, since in the actual polymer no similar shifts are found. The values calculated by Kunitake for the trans compound are 171.2 for the carbonyls and 43.1 ppm for carbons 2 and 3 of the ring. These are observed to be close to that found in the anhydride form of the copolymer, i.e., peak C at 43.6 ppm and peak G at 171.3 ppm.

We have obtained a commercial sample of the 2,3-dimethylsuccinic acid, a mixture of the racemic and meso forms. We have converted the acid to the anhydride and determined the ^{13}C spectrum without separation of the two isomers. The chemical shifts of the anhydride may be readily assigned from the peak intensities and the data of Ernst and Trowitzsch[5] and are shown in Figure 4. This is further supported by the fact that in the meso or cis form, the methyl carbons would be expected to be eclipsed and show a compression or steric shift. When we performed a calculation analogous to that of Kunitake as shown in Figure 5 but starting directly from the cis-2,3-dimethylsuccinic anhydride, we found that the predicted shifts for the cisoid form represented by the 2,3-diisopropyl structure were 33.5 and 171.0 ppm. Thus the carbonyl shift is in poor agreement with that calculated by Kunitake and is certainly in the range of that found for the actual polymer. On the strength of this evidence, we feel that cisoid backbone structures can not be ruled out using this type of argument.

The approach of extending data to account for major structure differences between a model and an actual compound is frequently used, and in fact, often valid. However, the above exercise

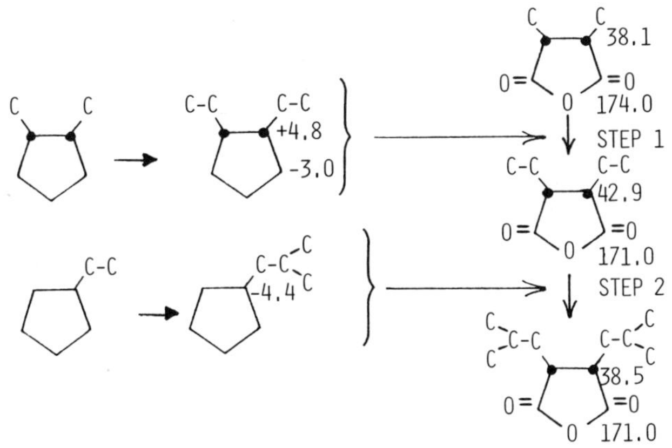

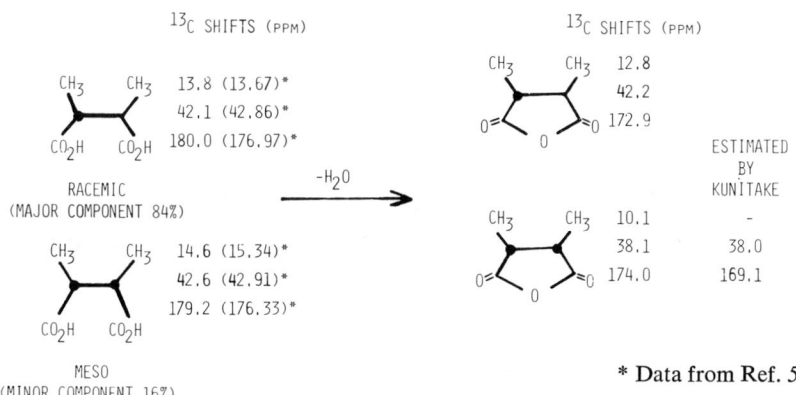

Figure 3. Scheme used by Kunitake (3) to estimate chemical shifts of succinate ring on copolymer backbone.

Figure 4. ^{13}C NMR data obtained for dimethyl succinic acid and anhydride. These data represent an improved model for estimating chemical shifts in a five-membered ring anhydride structure.

* Data from Ref. 5.

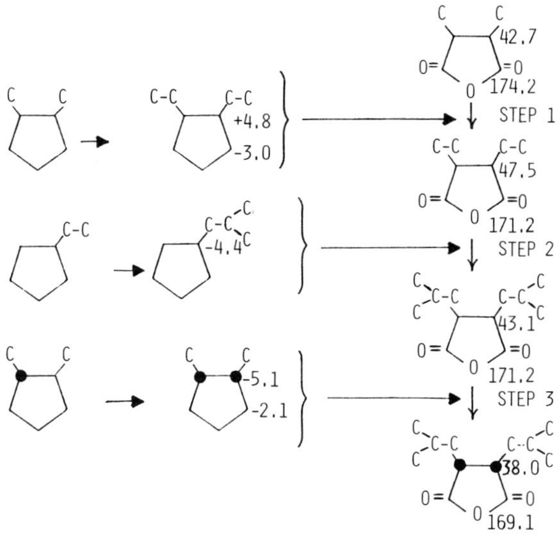

Figure 5. Improved procedure to estimate chemical shifts of succinate ring on copolymer backbone. This scheme is similar to that shown in Figure 3, but uses data from Figure 4. The results presented here show the danger in using such methods to make fine distinctions.

illustrates that such an approach is not without danger. Fine distinctions of the order of 1-2 ppm may be uncertain, and indeed unless one uses a model which is very close in structure, such calculations are likely to be misleading.

For example, Kunitake used the shifts on going from a trans- to a cis-1,2-dimethylcyclopentane as representative of that which would occur in an analogous 2,3-dimethylsuccinic anhydride. Thus he estimated a shift of -5.1 ppm for C-2 (and C-3) and -2.1 ppm for the carbonyl carbon. We found that the actual shifts on going from trans- to cis-2,3-dimethylsuccinic anhydride are -4.1 ppm (C-2 and C-3) and +1.1 ppm (carbonyls).

Five Membered Ring Model. The 2,5-dimethyl-3,4-tetrahydrofuran dicarboxylic acid was synthesized by means of the scheme shown in Figure 2b. The synthesis of this compound was complicated by the rather severe conditions necessary to accomplish the final hydrogenation step. These conditions were hydrogenation for 21 hours at 169-190°C at 3000 psi using a mixture of ruthenium on carbon and ruthenium chloride catalyst. This treatment gave rise to a mixture of isomers as well as an incompletely hydrogenated product. Nevertheless, the ^{13}C NMR spectra of the crude reaction product allowed the general assignment of the ring spectra of the 2,5-dimethyl-3,4-tetrahydrofuran to be readily made. These assignments are shown in Figure 2b. The ^{13}C shift of the ring carbons, C-2 and C-5, are of course of primary interest as models for the divinyl ether maleic anhydride copolymer. However, there is also a substantial interest in the sidechain carbons since it is through these shifts that the relative stereochemistry of the isomers is potentially established.

Now we could readily see from the ^{13}C NMR spectrum that there were three isomers present for this model compound. Upon fractionation of the mixture by gas chromatographic separation, we could establish that of the three isomers, two were symmetrical and one was nonsymmetrical by means of the ^{13}C NMR spectrum. These model compounds contain four asymmetrical centers. This in turn leads to a maximum of 2^N or 16 compounds where N=4. This, of course, represents a potential maximum of 8 dl pairs. Elimination of two possible isomers because of the additional symmetry due to having the groups on carbons 2 and 5 and the groups on carbon 3 and 4 identical, then leaves only six possible relative stereoisomers. Formally, this would be 5 dl pairs and one meso compound. This means, then, as a maximum or for the worst possible case, there are 20 possible configurations of 3 out of 6 isomers to consider. The figure 20 comes from the number of ways in which one may take 6 items 3 at a time.

During the course of a gas chromatographic workup of a mixture of the tetrahydrofuran diethylester we noticed that upon washing the crude mixture successively with NaOH, acid, and water that one of the GC peaks, F, decreased while one, A, grew. The remaining GC peaks remained approximately the same. These data

TABLE II

^{13}C AND PARTIAL ^{1}H NMR SHIFTS FOR STEREOISOMERS OF ETHYL 2,5-DIMETHYL-3,4-TETRAHYDROFURANDICARBOXYLATE

ISOMER	ELEMENT	CARBON NO. (PPM)								
		2	3	4	5	2,Me	5,Me	3,C=O	4,C=O	OCH_2-	$-CH_3$
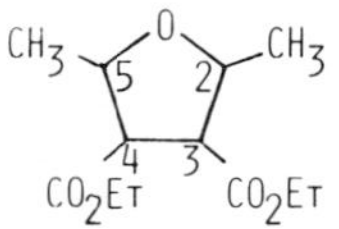 GC FRACTION E (SYMMETRICAL)	^{13}C	77.14	50.0	50.0	77.14	17.4	17.4	171.3	171.3	–	–
	^{1}H	4.60	3.64[A]	3.64[A]	4.60	–	–	–	–	60.5	14.2
GC FRACTION F (SYMMETRICAL)	^{13}C	75.1	51.0	51.0	75.1	17.2	17.2	170.4	170.4	–	–
	^{1}H	ca. 4.3	3.32[B]	3.32[B]	4.28	–	–	–	–	–	–
GC FRACTION A (UNSYMMETRICAL)	^{13}C	76.2	52.1[C]	54.0[C]	77.9	17.2	20.0	171.8	172.1	60.6	14.2
	^{1}H	–	3.13[D]	3.46[E]	–	–	–	–	–	–	–

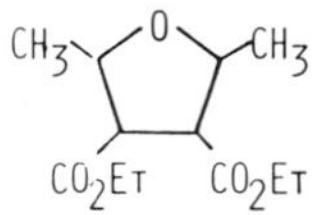

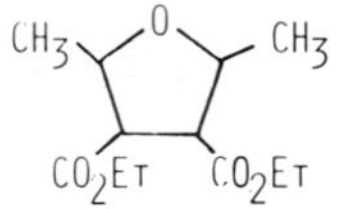

A) J = 2.07 AND 5.64 Hz
B) J = 1.91 AND 5.28 Hz
C) MAY BE REVERSED
D) J = 6.65 AND 8.39 Hz
E) J = 6.65 AND 7.8 Hz

indicate that one of the trapped isomers was interconverted to another upon treatment with either base or acid. This information suggests that an epimerization is occurring at the C-3 or C-4 carbon to interconvert one of the model isomers. These data, then substantially reduce the problem of characterizing mixtures of stereoisomers. When all the constraints or degrees of freedom are imposed upon the problem, we find that we still must consider 12 separate cases.

Conclusions

Unfortunately, our carbon-13 and proton data (Table II) taken together with other data for the model compounds, do not allow an unambiguous assignment of the 3 stereoisomers which we have obtained. For the present, then, we must be content to leave the relative stereochemistry of the 5-membered ring model compounds in question. For now, we can only say that the fine structure in the copolymer spectra are certainly well within isomers shifts seen for the model compound. However, based on the fact that the relative stereochemistry of the tetrahydrofuran models are unknown and that we have only one pyran isomer, we are unable to make a definite assignment as to the relative stereochemistry of ring junctions in the original copolymer.

Finally, our studies show all divinyl ether-maleic anhydride copolymers examined to date to be a mixture of the 5-membered and 6-membered bicyclic structures. The primary evidence for this, of course, is the presence of peak B at ca. 35 ppm which is assigned to the methylene carbon on a tetrahydropyran ring. Major features of the spectrum can be readily explained based on the available data present by both us and Professor Kunitake. However, still in question is the meaning of the fine structure contained particularly in peaks D, E, and F with respect to the relative stereochemistry of the copolymer as produced.

Acknowledgments

The Authors gratefully acknowledge the help and cooperation of Eleanor Medon and Carlene Eckroade in preparing the copolymer and model compounds.

Literature Cited

(1) G. B. Butler, J. Macromol Sci., Chem., A5, 219 (1971)
(2) D. S. Breslow, Pure & Appl. Chem., 46, 103 (1976)
(3) T. Kunitake and M. Tsukino, J. Polym. Sci., Poly., Chem. Ed., 17, 877 (1979).
(4) G. B. Butler and Y. C. Chu, ibid, 17 859 (1979).
(5) L. Ernst and W. Trowitzsch, Chem. Ber., 197, 3771 (1974)

RECEIVED July 14, 1981.

20

Interaction of Methotrexate (Polylysine) with Rat Liver Tumor Cells

JOHN M. WHITELEY
Scripps Clinic and Research Foundation, La Jolla, CA 92037

JOHN H. GALIVAN
New York State Department of Health, Division of Laboratories and Research, Albany, NY 12201

During the past decade interest has developed in the possible therapeutic use of carrier-bound drugs to secure greater interaction between the drug and its target site of action. Examples of carriers include liposomes (1), polypeptides (2), polysaccharides (3) and antibodies (4), and of drugs daunomycin (5), methotrexate (MTX) (6), actinomycin D and cytosine arabinoside (7). In this laboratory MTX, a potent inhibitor of one-carbon metabolism, coupled to albumin (BSA) has been shown to be equally effective to free MTX in the treatment of murine L1210 leukemia (8) and superior to the free drug in controlling the Lewis lung carcinoma (9). A recent examination of the interaction of this carrier-bound drug with L1210 cells in vitro has suggested that the nature of the carrier is important for interaction with the cell surface and that a carrier-bound drug might be absorbed into the cell by an alternate mechanism to that of the free agent (10). Because of the therapeutic significance of these observations, it was decided to investigate the interaction of MTX-BSA with a non-systemic cellular source such as the MTX transport-resistant Reuber H35 hepatoma cell line, which had been developed in one of the authors' laboratories (11). Unfortunately, in constrast to L1210 cells MTX-BSA was ineffective in supressing growth of the liver tumor cells. However, because of the observations that carrier structure can determine cellular interaction, and that polylysine possesses unique cellular absorptive properties (12), MTX was bound to both L- and D- forms of this carrier and the products were tested with the hepatoma cells in vitro. The comparative toxicities of the polymer-bound and free drugs with both parental and MTX transport-resistant cell lines were measured, and experimental results were obtained, which showed how differential cell rescue with reagents which usually overcome blockade of the one-carbon metabolic pathway could also afford similar results with cells treated with the drug-carrier complex.

Experimental

Materials. The following were obtained from commercial sources: Swims medium S77, horse serum, and fetal calf serum (Grand Island Biologicals); MTX (Lederle Laboratories) was purified by DEAE cellulose prior to use (11); crystalline BSA, poly-L-lysine ($M_r \sim 35,000$), hypoxanthine and thymidine (Sigma Chemical Co.); 1-ethyl-3-(3'-dimethylaminopropyl)carbodiimide hydrochloride (EDC) (Story Chemicals); folinic acid (ICN Pharmaceuticals); Blue Dextran (Pharmacia). The concentrations of the folate derivatives and MTX were determined by their respective extinction coefficients (13). MTX-BSA was synthesized according to a previously described procedure (8) in which MTX is coupled via a terminal carboxyl group to ϵ-amino groups contained in the albumin molecule.

MTX(poly-L- and poly-D-lysine) were synthesized according to a procedure first outlined by Ryser and Shen (14). 0.1 ml of a neutral aqueous solution of MTX (13 mg/ml) and 0.05 ml of an aqueous solution of EDC (200 mg/ml) were added to the polylysine samples ($\sim$ 10 mg) dissolved in 0.5 ml of 0.01 M potassium phosphate, pH 7.2, containing 0.15 M sodium chloride. After stirring in the dark at room temperature for 1.5 hours, the products were applied to columns of Sephadex G-25 (2.2 x 90 cm) which had been pre-equilibrated with the reaction buffer solution. Elution was carried out with the same buffer and the eluent absorbance was monitored at 280 nm. Fractions comprising the initial peak absorbance were combined and dialysed against water (6 liters) then lyophilized. The yellowish flocculent residues (wt. $\sim$7 mg) contained $\sim$5 mole MTX/mole polylysine when estimated by MTX absorbance at 370 nm ($\epsilon_{0.1\ N\ NaOH}$ = 7,100 (13)).

Cell culture. H-11-EC3 (designated H35 cells) derived from the Reuber H35 hepatoma (15,16) were grown as previously described and the development of resistant sublines was outlined in the same publication (11). MTX-resistant lines are indicated by the designation H35R with the subscript following the R indicating the concentration of MTX, in μM, to which the cells are resistant. All the cells employed in this study were resistant solely through defective transport and showed no more than 50% increase in dihydrofolate reductase relative to control cultures (21). Hepatocytes isolated from male rats of the Fisher strain by collagenase perfusion and mechanical dispersion were plated at a density of 7.5×10^5 cells/ml in L-15 medium containing 15% FBS and 10 mU/ml insulin on petri dishes according to a procedure outlined earlier by Tarentino and Galivan (17).

Growth Studies. The cells were cultured at a density of 5×10^4 cells/ml of Swims S77 medium with 20% horse serum and 5% fetal calf serum in a 95-5% air-CO_2 mixture at 37°C. The indicated additions were made at this time to duplicate samples and the cultures were allowed to incubate for 72 hours after which the cells were released from the plates with 0.05% trypsin and counted with a ZB1 Coulter counter (11).

Results

Treatment of H35 cells with MTX-BSA. The comparative growth rates of H35 cells plated at 2×10^5 cells/dish in the presence of various levels of MTX-BSA are shown in Table I. Clearly the growth of wild-type cells is retarded by micromolar quantities of MTX-BSA with 40 μM MTX-BSA leading to a complete cessation of growth. The $H35R_1$ (see EXPERIMENTAL for subscript definition) cells, however, are unaffected by this concentration of carrier-bound drug and demonstrate a growth response comparable to that of the untreated H35 cells.

Treatment of H35 cells with MTX(poly-L-lysine). The influence of various concentrations of MTX(poly-L-lysine) on both H35 and $H35R_{0.3}$ hepatoma cells is illustrated in Figure 1. The effect is also compared to that of the free drug. It is apparent that MTX shows little restraint on the growth of the transport-resistant mutants at levels up to 1 μM in contrast to the wild-type cells which submit to nanomolar concentrations of the drug (I_{50} = 14 nM). However, when MTX is bound to the poly-L-lysine carrier, both cell lines show similar characteristics with I_{50} ~70 nM. The concentrations of poly-L-lysine used in the experiments show little influence on the cell growth.

Prevention of MTX and MTX(poly-L-lysine) toxicity to H35 cells. Table II shows the differential effects which are observed when H35 and $H35R_1$ cells treated with MTX (0.1 μM) and MTX(poly-L-lysine) (1.0 μM) are rescued with folinic acid (100 μM) or thymidine/hypoxanthine (500/100 μM). The H35 cells are sensitive to each drug and respond in the case of MTX and MTX(poly-L-lysine) to supplementation with folinic acid although those treated with the free drug give a better response than those treated with the carrier-bound agent (92% as opposed to 69% of normal growth rate). The $H35R_1$ cells, unaffected by MTX, but susceptible to MTX(poly-L-lysine) do not respond to folinic acid. However, both cell types show a moderate response to thymidine/hypoxanthine rescue.

TABLE I: The comparative growth of resistant and non-resistant H35 cells treated with MTX-BSA.

Cell type	Concentration MTX-BSA (µM)	Cells/plate at 72 hr
H35	0	2.7×10^6
H35	10	9.4×10^4
H35	40	7.3×10^4
H35R	0	2.2×10^6
H35R	10	2.2×10^6
H35R	40	1.8×10^6

TABLE II: The effect of folinic acid and thymidine/hypoxanthine on H35 and $H35R_1$ cells treated with MTX(poly-L-lysine) and MTX.

MTX (poly-L-lysine) (1.0 µM)	MTX (0.1 µM)	Folinic Acid (100 µM)	Thymidine/ Hypoxanthine (500 µM/100 µM)	Cells[1] (%)
H35				
+				9
+		+		60
		+		107
+			+	36
			+	78
	+			6
	+	+		92
$H35R_1$				
+				15
+		+		20
		+		98
+			+	40
			+	87

[1]Number of cells are expressed relative to the untreated cultures.

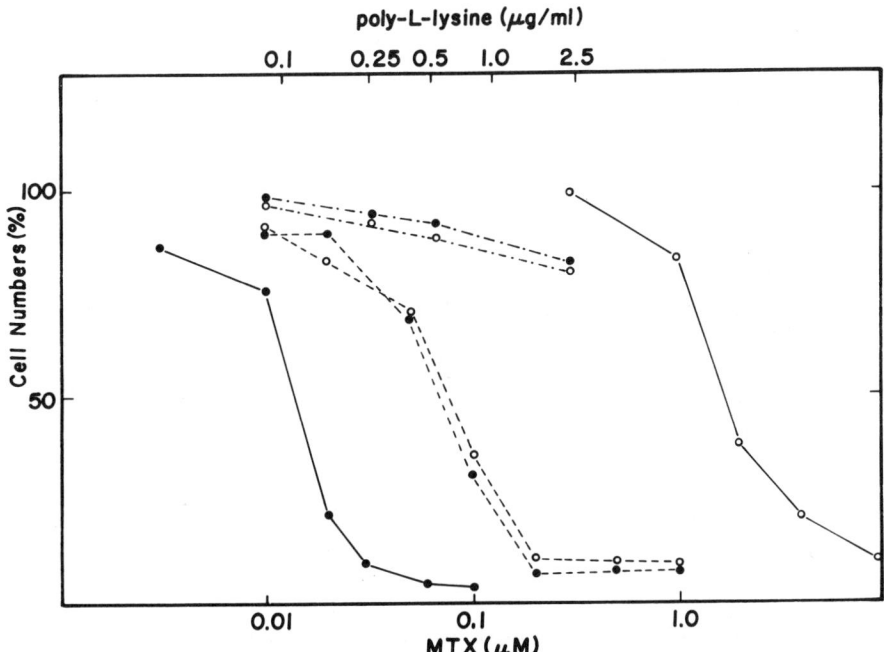

Figure 1. Comparative growth curves of resistant (○) $H35R_{0.3}$ and sensitive (●) H35 cells when treated with MTX (———), MTX(poly-L-lysine) (– – –), and (poly-L-lysine) (–●–) alone at the levels indicated. The growth rates are expressed as a percentage of those observed with untreated cell samples. (Reproduced, with permission, from Ref. 23. Copyright 1981, American Society for Pharmacology and Experimental Therapeutics.)

Measurement of dihydrofolate reductase and thymidylate synthetase activities in MTX and MTX(poly-L-lysine) treated H35 cells. Extracts of the untreated cells were prepared from two 60 mm plates, and four 100 mm plates were used for the extracts of inhibited cells. Dihydrofolate reductase activity was measured by a modification (18) of the method of Mathews (19) and thymidylate synthetase in the intact cell by the release of ^{3}H from [5-^{3}H] deoxyuridine as described previously (20). When either MTX transport-defective or nondefective cells were treated with MTX and MTX(poly-L-lysine), it was found that both drugs, administered at the concentrations shown in Figure 2, led to elimination of both dihydrofolate reductase and thymidylate synthetase activities.

Sephadex chromatography of $H35R_{0.3}$ cell extracts after exposure to [^{3}H]MTX(poly-L- and poly-D-lysine). After 72 hours in culture thirty 60 mm plates of H35 cells were exposed for a further 24 hours to either [^{3}H]MTX(poly-L- or -D-lysine) (0.75 mM, 3 x 10^5-dpm/nmol). At the end of the incubation the plates were placed on ice and washed x 4 with 4 ml ice-cold Hanks balanced salt solution then scraped into two 1 ml aliquots of isotonic saline, buffered with 0.01 M potassium phosphate, pH 7.4. The cells were broken by 20 strokes with a Dounce homogenizer and centrifuged at 15,000 x g for 20 min. 1 mg of Blue Dextran 200 (M_r = 2 x 10^6) was added to the extracts and each of the mixtures (poly-L- or poly-D-) was chromatographed, as described previously (20), on Sephadex G-75 (Figure 3). When mixtures of [^{3}H](poly-L- or -D-lysine) and Blue Dextran 2000 were mixed in separate experiments and applied to the same column, radioactivity eluted only in the void volume corresponding to the first peak. In the MTX(poly-L-lysine) experiment the second peak was identified by its enzymatic activity as dihydrofolate reductase and the third peak corresponded to a mixture of higher molecular weight γ-glutamyl conjugates of MTX and free MTX.

When the contents of peak 2 were boiled at pH 8.0 and applied to a second column of Sephadex G-25 the profile observed in Figure 3A was obtained whose peaks corresponded to known standards for free MTX and polyglutamate forms of MTX. Similar results were obtained when an aliquot of combined fractions 55 to 77 were applied to the Sephadex G-25 column (Figure 3B).

In contrast when the cell extracts from the [^{3}H]MTX(poly-D-lysine) experiment were applied to the Sephadex G-75 column only the undegraded complex with Blue Dextran 200 was observed. When the experiment was repeated with the non-resistant H35 cells, similar results were obtained.

Uptake of MTX and MTX(poly-L-lysine) by hepatocytes, H35 and $H35R_1$ cells in culture. Hepatocytes or hepatoma cells maintained in culture (11) were treated with [^{3}H]MTX(poly-L-lysine) or [^{3}H]MTX. At the indicated times (Figures 4 and 5) cells were washed, and the monolayer was measured for radioactivity (11). Clearly higher concentrations of carrier-bound rather than free drug remained associated with the cells in every case.

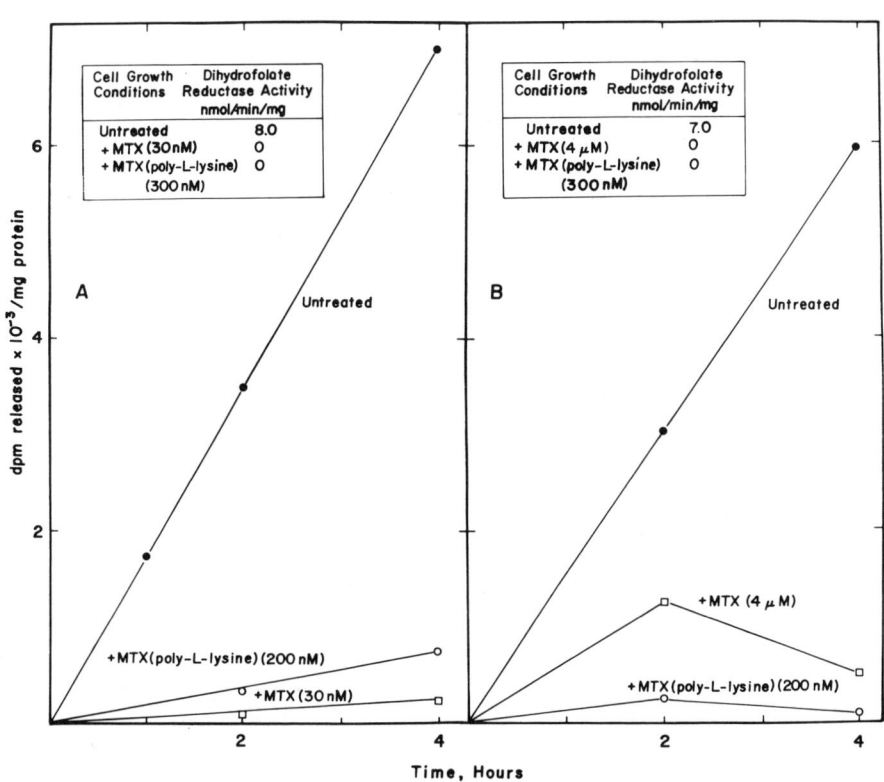

Figure 2. Inhibition of dihydrofolate reductase and de novo thymidylate biosynthesis in (A) nonresistant H35 and (B) resistant $H35R_{0.3}$ hepatoma cells, after treatment with MTX and MTX(poly-L-lysine). (Reproduced, with permission, from Ref. 23. Copyright 1981, American Society for Pharmacology and Experimental Therapeutics.)

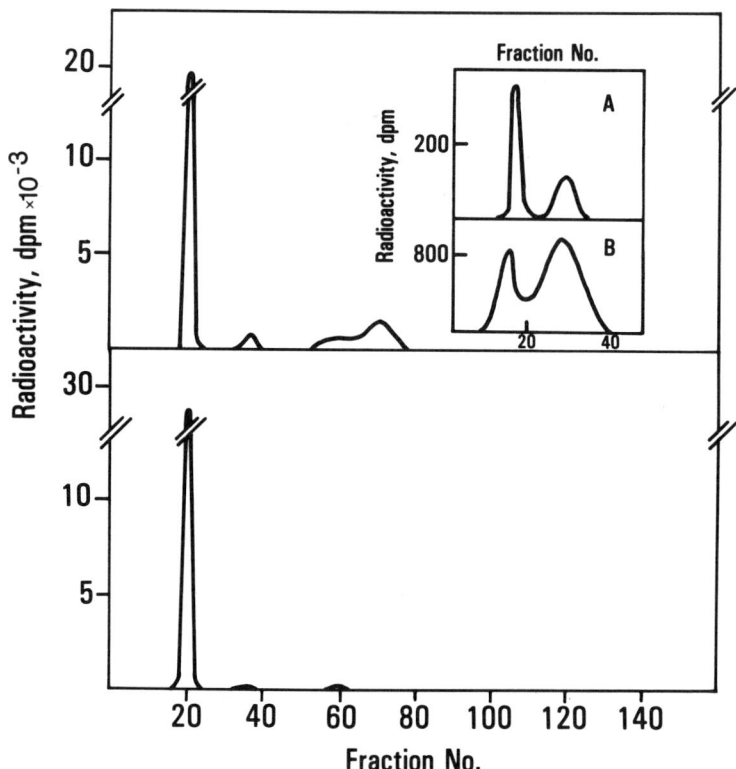

Figure 3. Sephadex G-75 chromatography of $H35R_{0.3}$ cell extracts after exposure to [^{3}H]MTX(poly-L-lysine) (upper panel) and [^{3}H]MTX(poly-D-lysine) (lower panel). Key: A, rechromatography of Sephadex G-25 of the contents of the second peak after boiling at pH 8.0 and B, chromatography on Sephadex G-25 of an aliquot from the combined fractions 55 to 77.

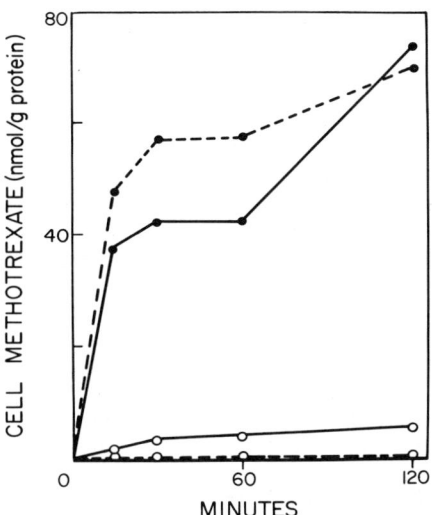

Figure 4. Uptake of [³H]MTX (○) and [³H]MTX(poly-L-lysine) (●) by H35 (———) and H35R₁ (– – –) cells in culture. The medium concentration of MTX in each case was 0.3 μM.

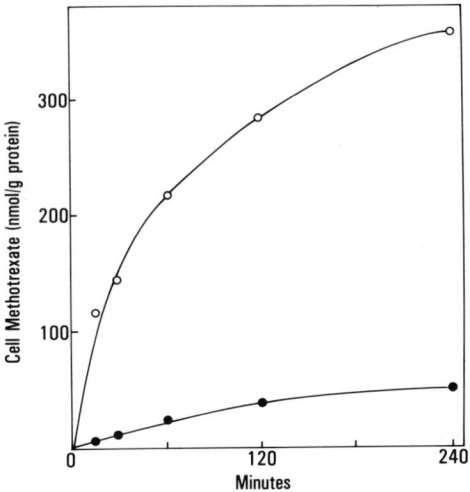

Figure 5. Uptake of [³H]MTX (●) and [³H]MTX(poly-L-lysine) (○) by hepatocytes in culture. The medium concentration of MTX in each case was 1 μM.

Toxicity of poly-L- and -D-lysine derivatives to rat hepatocytes, H35 and H35 $R_{0.3}$ cells in culture. Cultured hepatocytes or hepatoma cells were treated with differing levels of either poly-L- or poly-D-lysine. The effect of these agents on cell viability is shown in Figures 6 and 7. It is apparent that both D- and L-polymers show limited toxicity towards the hepatoma cell lines, but that the D-isomer shows considerable toxicity towards the hepatocytes. Figure 7 also shows that the MTX derivatives of both D- and L-carriers are equally toxic to the non-substituted agent when interacting with the hepatocytes. It should also be noted that MTX linked to poly-D-lysine does not enhance the toxicity of the carrier to H35 cells.

Effect of MTX(poly-D- and poly-L-lysine) on rat hepatocyte dihydrofolate reductase. Hepatocytes were maintained in culture for 48 hours prior to measurement of viability and dihydrofolate reductase activity (Table III). Even at concentrations up to 2 μM MTX(poly-D-lysine) causes little inhibition of cellular dihydrofolate reductase, however, it shows considerable cellular toxicity. In contrast MTX(poly-L-lysine) at these levels maintains good cellular viability with somewhat higher but not complete inhibition of the reductase. This clearly demonstrates that the toxicity of MTX(poly-D-lysine) is exerted through the carrier by some, as yet, unidentified mechanism and that little MTX is released to interact with the reductase. The comparative figures for treatment with free MTX indicate complete inhibition of enzyme activity. Since normal hepatocytes consist of a non-dividing population, extensive inhibition of dihydrofolate reductase might not be expected to interfere with cell viability.

Effect of MTX(poly-L-lysine) pulse doses on hepatoma cell growth and dihydrofolate reductase. Cells were maintained in culture for 72 hours prior to measurement and the pulse dose was added between 24 and 48 hours. The results are shown in Table IV where it is apparent that 2.8 μM MTX(poly-L-lysine) is effective in reducing cell growth of both defective and non-defective cell lines, and also eliminates dihydrofolate reductase activity.

Discussion

The results described in this report amplify observations that carrier structure is integral to the concept of polymer-mediated drug delivery (10). Two general features begin to emerge from our studies on carrier-bound drug therapy: (1) that in a whole animal experiment the carrier-bound agent ensures elevated levels of the drug are retained compartmentalized within a body cavity thus providing a source of continous high level drug delivery (8,9), and (2) that polymer-bound drugs can overcome transport blocks which certain cells use to prevent passage of the free drug into the cell. Because earlier studies had shown the effectiveness of MTX-BSA in the treatment of certain mouse tumors (8,9), we considered extend-

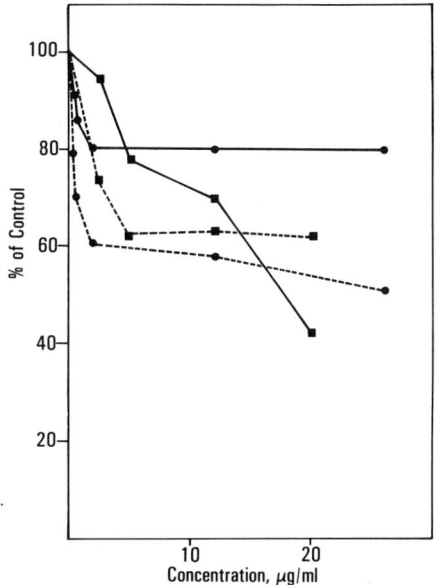

Figure 6. Toxicity of poly-D-lysine (■) and poly-L-lysine (●) against H35 (———) and $H35R_{0.3}$ (- - -) rat hepatoma cell lines.

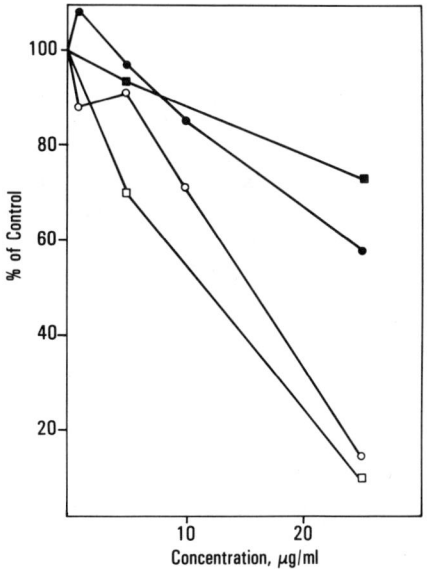

Figure 7. Toxicity of poly-L-lysine (●), MTX(poly-L-lysine) (■), poly-D-lysine (○), and MTX(poly-D-lysine) (□) to rat hepatocytes in culture.

TABLE III: Effect of MTX(poly-D-lysine) and MTX(poly-L-lysine) on rat hepatocyte dihydrofolate reductase.

DRUG ADDED	CONCENTRATION MTX, polylysine		DHFR	CELL VIABILITY
	μM	μg/ml	%	%
Control			100	100
MTX-poly-L-lysine	0.2	2.5	58	95
	2.0	25	26	75
MTX-poly-D-lysine	0.2	2.5	80	70
	2.0	25	74	10
MTX	0.2		10	100
MTX	2.0		0	92

Measurements were made after 48 hours in culture with drug addition after 24 hours.

ing its application to the growth control of a chemically induced hepatoma cell line and its transport-resistant counterpart. The reagent, however, proved ineffective and alternate carriers had to be considered. The demonstrated toxicity of MTX(poly-L-lysine) against Chinese hamster ovary cells (14) suggested this amino-acid polymer as a possible effective delivery system. The results demonstrate that MTX(poly-L-lysine) is also effective in this case and that it bypasses the transport defect, as was also recently observed by Shen and Ryser in L929 mouse fibroblasts (24). MTX-(poly-L-lysine), however, is slightly less effective than MTX against the parent cell line. It has also been shown that the cells subject to chemotherapy with the MTX(polylysine) complexes can be rescued with hypoxanthine/thymidine but only the non-resistant cells can be rescued with folinic acid. The observation that folinic acid does not reverse toxicity in the resistant cells is consistent with the transport mutants not being able to take up either MTX or the reduced folates (21) as has already been shown with MTX or pyrimethamine toxicity in MTX transport-resistant L-5178 lymphoblasts (22). Although MTX(poly-D-lysine) is toxic to H35 cells and hepatocytes, it is no more so than poly-D-lysine itself. The experiments in which the lysates of hepatoma cells exposed to MTX(poly-L and -D-lysine) derivatives are examined by Sephadex chromatography (Figure 3) amplify this observation. Only with the L-isomeric carrier are both free MTX and enzyme-bound drug found in the cytoplasm. With MTX(poly-D-lysine) only an unchanged high-molecular weight species can be identified.

Experiments which examine the comparative toxicities of the poly-L- and -D- carriers for hepatocytes and hepatomas indicate a low level of toxicity with the hepatoma. However, an elevated toxic effect is observed for the hepatocyte with the D-isomeric carrier. When the drug is bound to the carrier little alteration is observed in the toxic response of the hepatocyte suggesting that in its resting state it has little requirement for dihydrofolate reductase. The results shown in Table III illustrate this point more clearly where only 26% of reductase activity remains after treating the hepatocytes with 2 micromolar MTX(poly-L-lysine) yet 75% cellular viability remains. This contrasts with the response of the hepatoma cell line (Table IV), where both dihydrofolate reductase and ~90% cell growth of both transport defective and non-defective cell lines is eliminated by treatment with 2.8 micromolar MTX(poly-L-lysine).

Although <u>in vitro</u> experiments do not necessarily predicate successful experiments in the whole animal the differential response of the hepatocytes and hepatoma cells to the MTX(poly-L-lysine) suggests that treatment of rats bearing similar liver tumors could be successful.

TABLE IV: Effect of a pulse dose of MTX(poly-L-lysine) on hepatoma cell growth and dihydrofolate reductase activity.

CELL TYPE	DRUG ADDITION	MTX CONC^N	DHFR	CELL GROWTH
		μM	%	%
H35			100	100
	MTX-poly-L-lysine	2.8[a]	0.3	7.6
H35R$_{0.3}$			100	100
	MTX-poly-L-lysine	2.8[a]	0.1	11

Cells were cultured for 72 hours prior to measurement.

[a]Pulse dose added between 24 and 48 hours.

Legend of Symbols

MTX, Methotrexate; BSA, bovine serum albumin; EDC, 1-ethyl-3(3'-dimethylaminopropyl)carbodiimide hydrochloride.

Acknowledgments

The work was supported by USPHS grants CA11778, CA25933 and AG00207.

Literature Cited

1. Gregoriadis, G. New Engl. J. Med. 1976, 295, 704-710 and 765-770.
2. Barbanti-Brodano, G. and Fiume, L. Nature 1973, 243, 289-283.
3. Harding, N.G.L. Ann. N.Y. Acad. Sci. 1971, 186, 270-283.
4. Rowland, G.F. Europ. J. Cancer 1977, 13, 593-596.
5. Hurwitz, E.; Maron, R.; Bernstein, A.; Wilchek, M.; Sela, M. and Arnon, R. Int. J. Cancer 1978, 21, 747-755.
6. Fung, W.-P.; Przybylski, M.; Ringsdorf, H. and Zaharko, D.S. J. Nat. Cancer Inst. 1979, 62, 1261-1264.
7. Szekerke, M. and Driscoll, J.S. Europ. J. Cancer 1977, 13, 529-537.
8. Chu, B.C.F. and Whiteley, J.M. Mol. Pharmacol. 1977, 13, 80-88.
9. Chu, B.C.F. and Whiteley, J.M. J. Natl. Cancer Inst., 1979, 62, 79-82.
10. Chu, B.C.F. and Whiteley, J.M. Mol. Pharmacol. 1980, 17, 382-387.
11. Galivan, J.H. Cancer Res. 1979, 39, 735-743.
12. Shen, W.-C. and Ryser, H.J.-P. Proc. Natl. Acad. Sci. U.S.A. 1978, 75, 1872-1876.
13. Blakely, R.L. "The Biochemistry of Folic Acid and Related Pteridines" (Eds., A. Neuberger and E.L. Tatum); North Holland Publ. Co.: Amsterdam and London, 1969, 92-94.
14. Ryser, H.J.-P. and Shen, W.-C. Proc. Natl. Acad. Sci. U.S.A. 1978, 75, 3867-3870.
15. Pito, H.; Periano, C.; Moose, P. and Potter, V.R. Natl. Cancer Inst. Monograph 1964, 13, 229-245.
16. Reuber, M.D. J. Natl. Cancer Inst. 1964, 26, 891-899.
17. Tarentino, A.L. and Galivan, J. In Vitro 1980, 16, 833-846.
18. Bonney, R.J. and Maley, F. Cancer Res. 1975, 35, 1950-1956.
19. Mathews, C.K. J. Biol. Chem. 1967, 242. 4083-4086.
20. Galivan, J. Mol. Pharmacol. 1980, 17, 105-110.
21. Galivan, J.H. Cancer Res. 1981, 41, 000-000.
22. Goldie, J.H.; Harrison, S.I.; Price, L.A. and Hill, B.T. Eur. J. Cancer 1975, 11, 627-632.
23. Whiteley, J.M.; Nimec, Z. and Galivan, J. Mol. Pharmacol. 1981, 19, 000-000.
24. Shen, W.-C. and Ryser, H.J.-P. Life Sci. 1981, 28, 1209-1214.

RECEIVED July 14, 1981.

Evidence of Cooperativity and Allosterism in the Binding of Various Antibiotics and Carcinogens to DNA

L. S. ROSENBERG, M. S. BALAKRISHNAN, D. E. GRAVES, K. R. LEE, S. A. WINKLE, and T. R. KRUGH

University of Rochester, Department of Chemistry, Rochester, NY 14627

The interaction of a number of drugs (actinomycin D, adriamycin, daunorubicin, and DHAQ) and carcinogens (NQO and HAAF) with DNA is studied by optical titration and/or phase partition. In each case pronounced curvature of the Scatchard plots is observed at low levels of bound drug, indicative of a cooperative binding process. The overall shapes of the binding isotherms are also consistant with a drug induced allosteric change in the solution structure of the DNA. Salt concentration influences the magnitude and curvature of the binding isotherms. The addition of small concentrations of daunorubicin affects the shape of the actinomycin D binding isotherm, in a manner which suggests that both drugs are able to induce similiar allosteric changes in the DNA. The dissociation kinetics of actinomycin D from calf thymus DNA is shown to fit a multi-exponential decay curve, whereas the dissociation from poly(dGdC)·poly(dGdC) shows a single lifetime. The dissociation of the ternary complex actinomycin D:daunorubicin: poly(dAdT)·poly(dAdT) is best described by two time constants, while the dissociation of daunorubicin from the same polymer in the absence of actinomycin D is characterized by a single time constant.

During recent years a host of solid state and solution experiments have shown that DNA may exist in a variety of conformations (e.g., 1-10 and references therein). Factors affecting the solution conformation of DNA include salt concentration (7,8,9), nature of the solvent (10) and base content (i.e. AT/GC ratio)(3). Pohl et al. (7) showed that at high salt concentrations the solution structure of the heteropolymer poly(dGdC)·poly(dGdC) is altered to the so-called "Z" conformation or left-handed helix, as evidenced by an inversion of the circular dichroism spectrum. The binding of ethidium to the high salt form of the polymer is a markedly cooperative process (7). Ethidium binding to the high-salt conformation of this polynucleotide alters its conformation to resemble the low salt ethidium: polynucleotide complex.

Crothers and coworkers (1,11) have shown that the binding of the oligopeptides netropsin and distamycin to calf thymus DNA is a cooperative process. These results have been interpreted in terms of a drug induced allosteric change in the conformation of the DNA. This report shows that several important antitumor drugs whose chemical structures are shown in Figure 1 (i.e. actinomycin D, adriamycin, daunorubicin, DHAQ) and two carcinogens (NQO and HAAF) bind cooperatively to DNAs. Factors affecting the solution structure of DNA are shown to alter the degree of cooperativity. The presence of cooperativity and/or allosteric transitions in the binding of these ligands to DNA may be an important aspect of their pharmacological activity.

Crothers and coworkers (1,11) have also shown that the binding of ethidium to calf thymus DNA in the presence of distamycin (r(distamycin) = 0.037) is cooperative (1), whereas the binding of ethidium in the absence of distamycin is non-cooperative. We report that the binding of daunorubicin to calf thymus DNA (r(daunorubicin) = 0.037) appears to alter the structure of the DNA as evidenced by the non-cooperative isotherm of actinomycin D.

Müller and Crothers (12) have shown that the dissociation of actinomycin D from DNA is a multistep process. The relaxation curve for the dissociation of actinomycin D from calf thymus DNA requires a minimum of three exponentials to fit the decay curve mathematically. They associated the multistep dissociation kinetics to the presence of the pentapeptide rings of actinomycin D which are essential for the pharmacological activity of the drug. Recently, Krugh et al. (13) showed that the dissociation of actinomycin D from poly(dGdC)·poly(dGdC) was characterized by a single exponential decay. This and other related experiments have led Krugh and coworkers (unpublished results) to conclude that the different dissociation lifetimes on native DNA result from the dissociation of actinomycin D from different DNA binding sites. Presented below is a detailed examination of the dissociation of actinomycin D from both calf thymus DNA and poly(dGdC).poly(dGdC). Krugh and Young (14) have shown that daunorubicin facilitates the

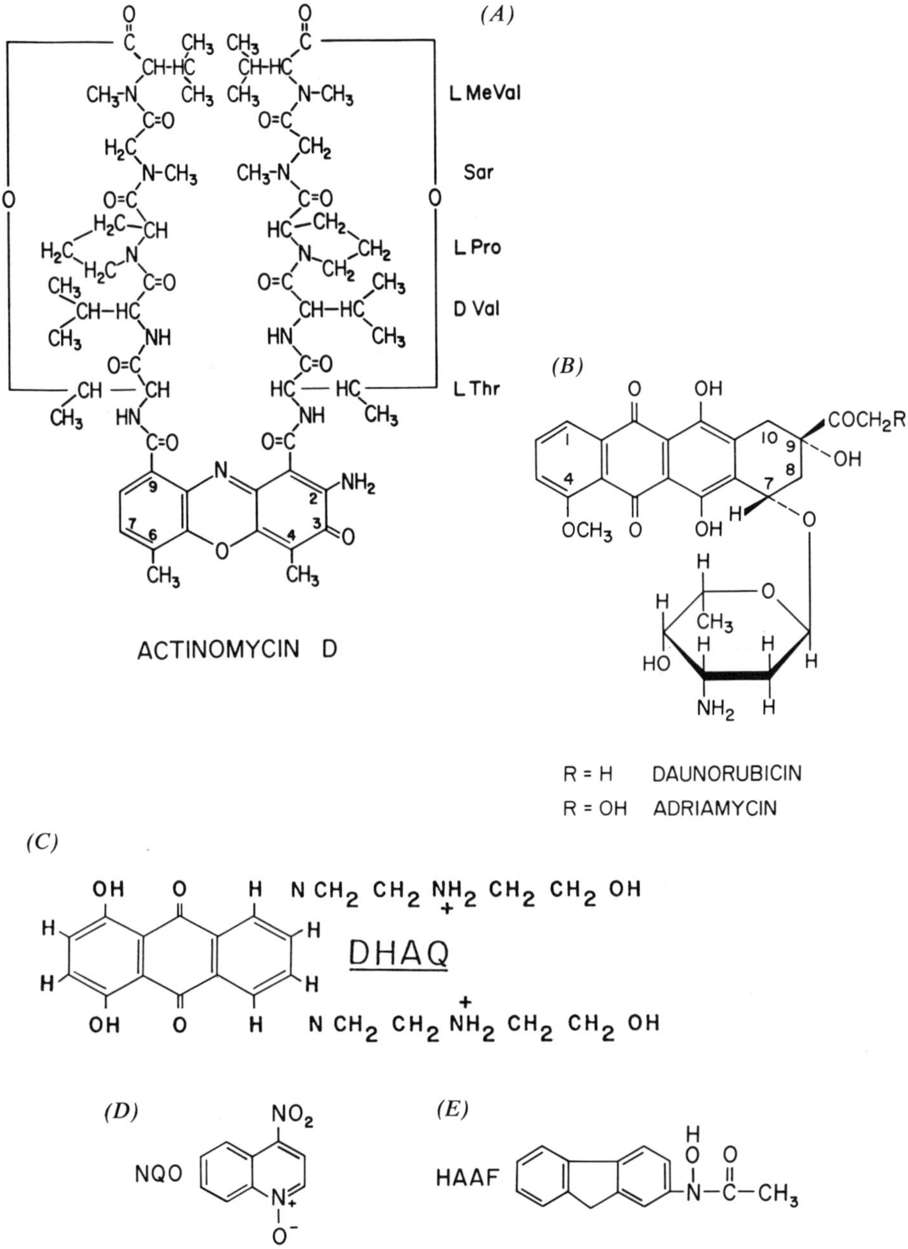

Figure 1. Chemical structure of (A) actinomycin D (key: Thr, threonine; Val, valine; Pro, proline; Sar, sarcosine; and MeVal, methylvaline); (B) adriamycin and daunorubicin; (C) DHAQ; (D) NQO; and (E) HAAF.

binding of actinomycin D to poly(dAdT)·poly(dAdT), a polymer to which actinomycin D binds only weakly. This paper shows that the dissociation kinetics of daunorubicin, in the presence and absence of actinomycin D, reflects the synergistic interaction of these two drugs as mediated by poly(dAdT)·poly(dAdT).

Materials and Methods

1,4-dihydroxy-5,8-bis[[2-[(2-hydroxyethyl)amino]ethyl]amino]-9,10-anthracenedione (DHAQ) a gift from Dr. H. Cox (University of Kansas Medical Center). Actinomycin D was obtained from Merck. Daunorubicin and adriamycin were from the Natural Products Branch, National Cancer Institute. ^{3}H-NQO (4-nitroquinoline-1-oxide) was from the laboratory of Dr. I. Tinoco, Jr. (U.C. Berkeley), ^{14}C-HAAF (N-Hydroxy-N-acetyl-2-amino-fluorene was from the ICN Radiochemicals Division, and ^{3}H-actinomycin D was from Amersham. Calf thymus DNA (Worthington), was prepared according to the procedure of Müller and Crothers (15). Poly(dAdT)·poly(dAdT) and poly(dGdC)·poly(dGdC) were purchased from PL Biochemicals. DNA concentrations were determined absorptiometrically, and expressed in basepairs.

Free and bound concentrations of DHAQ were determined by optical titration using a 10-cm quartz cell. A known volume and concentration of DNA was titrated with a concentrated solution of DHAQ. The increase in the absorbance was read directly from the digital display of a Cary 219 spectrophotometer. Buffer solutions were pH=6.4, 0.052M Na phosphate, 0.1M NaCl, and 0.001M NaEDTA.

The binding of actinomycin D, adriamycin, daunorubicin, HAAF and NQO to nucleic acids was examined utilizing the phase partition technique which allows the bound and free drug concentrations to be determined by direct measurements. Phase partition analysis (16-19, and references therein) consists of the equilibration of an aqueous phase (with or without DNA present) with an immiscible organic phase. The drug may be initially present in either phase. The partition coefficient, P (drug concentration in organic phase/drug concentration in aqueous phase) is determined in the absence of DNA. Upon addition of DNA to the aqueous phase, both phases are equilibrated and seperated. The concentration of the drug in the organic phase is measured and used to calculate the free drug concentration (cf) = [organic drug concentration]/ partition coefficient, P. Drug concentrations in both the organic and aqueous phase can be calculated directly by either scintillation counting or spectrophotometric measurements, as appropriate.

Binding isotherms for both adriamycin and daunorubicin were obtained by phase partition analysis utilyzing both 1-nonanol and chloroform as the organic phase. Aqueous buffer solutions were pH=7, 0.01M Na phosphate, 0.1M NaCl, and 0.001M NaEDTA. Aqueous and organic phases were equilibrated directly in teflon stoppered fluorimeter cells. Organic phase drug concentrations were determined by direct spectrofluorimetric measurements. Bound drug concentrations were calculated from the total drug concentration in the aqueous phase obtained spectrofluorimetrically after dissociation of the DNA-drug complex by the addition of dimethylsulfoxide (1:1 v/v).

The organic phase for radioactively labeled actinomycin D, NQO, and HAAF was a 1:1 mixture of n-chloroheptane and cyclohexane. For the experiments with HAAF and NQO the aqueous phase was a buffer of 0.1M NaCl, 0.01M Na cacodylate, 0.001M NaEDTA, pH=7. For actinomycin D the buffer was 0.01 M Na phosphate, 0.001M NaEDTA, pH=7, with the addition of NaCl as specified in the Results and Discussion section.

Actinomycin D dissociation kinetics were measured on a Cary 219 spectrophotometer equipped with a magnetic stirrer and thermostated cell holders. Sodium dodecyl sulfate (SDS) was used to sequester dissociating actinomycin D, and the resulting increase in absorbance was monitored at 452 nm as a function of time. Stop-flow studies (daunorubicin and daunorubicin/ actinomycin D) were conducted with a Durrum-Gibson Model 110 stopped-flow spectrophotometer equipped with a dual detector accessory and a Tektronix storage oscilloscope interfaced with a PDP 11/34 computer. Experiments were done in a 0.01M Na phosphate buffer, 0.1M NaCl, 0.001M NaEDTA, pH=7. Dissociation time constants were computed with a multiexponential analysis computer program.

Results and Discussion

Cooperative Binding. The binding isotherm for the interaction of DHAQ with calf thymus DNA (Figure 2) shows an initially increasing slope which is indicative of cooperative binding of DHAQ. At higher r values the binding isotherm changes, and appears to reflect a typical neighbor exclusion binding isotherm.

For DHAQ, increasing either the ionic strength of the buffer (0.2 (0.10M NaCl) to 0.6 (0.50M NaCl)) or the temperature (23 °C to 37 °C) diminishes the initial increasing slope of the DHAQ/DNA binding isotherm. At an ionic strength of 0.6, the slope is only slightly positive in the initial portion of the binding isotherm. Increasing the temperature, at any ionic strength, decreases the initial slope and shows a minimally decreasing Scatchard plot at low concentrations of bound DHAQ. Increasing the pH from 6.4 to 7.1 does not affect the shape of the binding isotherm at low ionic strength. Also, saturating the buffer with 1-nonanol (an organic solvent used in partition analysis experiments) does not affect the shape or magnitude of the binding isotherm.

The binding isotherms for the interaction of the anthracycline antibiotics, adriamycin and daunorubicin, with calf thymus DNA, measured by phase partition techniques, are shown in Figure 3. Both drugs show initially increasing binding isotherms, indicative of a cooperative binding process, and reach a maximum

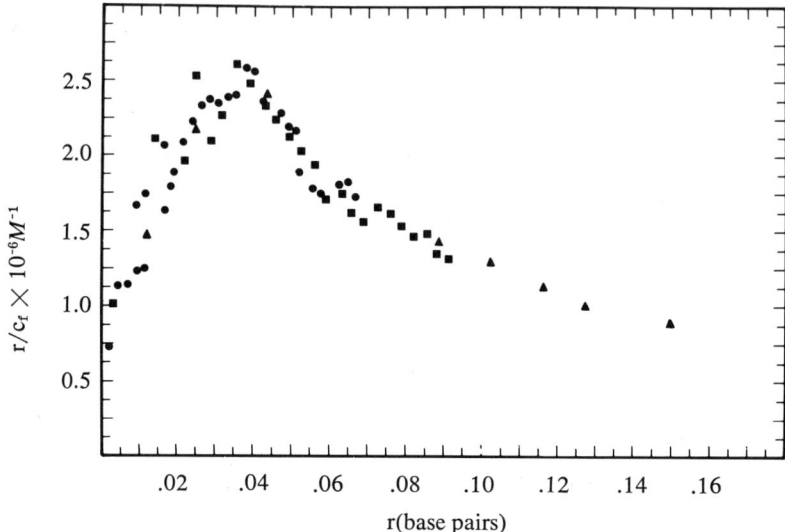

Figure 2. The binding of DHAQ by calf thymus DNA was evaluated by absorption spectroscopy.

The buffer was sodium phosphate (0.052M), sodium chloride (0.10M), and NaEDTA (0.001M) at pH 6.4 and final ionic strength 0.201. The wavelength of choice was the free drug maxima of 660 nm. Concentrations of the DNA (base pairs) ranged from 7 to 8 × 10^{-6}M, while the drug concentration ranged from 0.3 to 8.0 × 10^{-7}M.

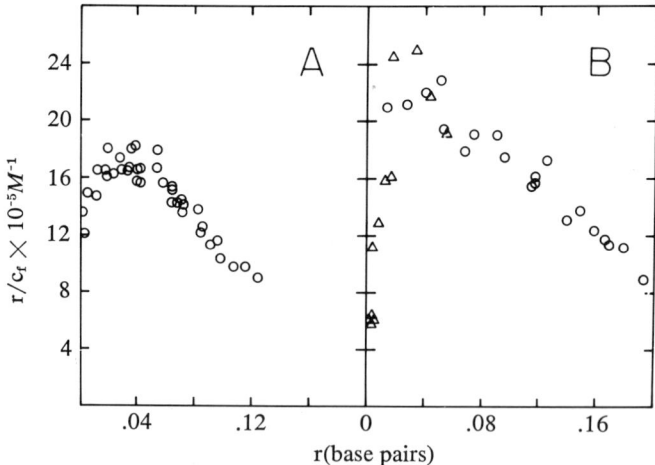

Figure 3. Scatchard plot of the equilibrium binding of calf DNA with (A) daunorubicin and (B) adriamycin at 22°C in 0.01M sodium phosphate, 0.1M sodium chloride, and 0.001M NaEDTA at pH 7.0.

Key: ○ and △, binding isotherms obtained using 1-nonanol and chloroform, respectively. Bound and free drug concentrations were determined by spectrofluorometric measurements, and excitation and emission wavelengths were 475 and 595 nm, respectively. Drug concentrations ranged from 0.1 to 0.7 μM, while DNA concentrations ranged from 0.15 to 15μM (base pairs).

at r = 0.035-0.040. The cooperativity of both drugs diminish (loss of the initial positive slope) as the sodium chloride concentration (0.10M - 0.01M) is decreased, thus indicating that cooperativity is dependent on the ionic strength.

The binding of actinomycin D to calf thymus DNA (0.01M, 0.1M, and 0.2M NaCl) and to poly(dGdC)·poly(dGdC) (0.1M NaCl) (Figure 4) was also examined by phase partition techniques. For both DNAs, a cooperative process was seen. In the case of calf thymus DNA, the decreasing salt concentration appears to increase the degree of cooperativity.

The effects of salt concentration on the binding of various drugs to DNA have been well documented at higher r values (21-24). Studies by Record et al. (25) and Manning (26) have shown that the counterions of DNAs are "obligate participants" in the interaction of charged ligands with DNA. Although the observed magnitude of the binding isotherms obtained is decreased with increasing [Na^+], the effect of salt on the binding at low r values, where the cooperative binding and allosteric transitions are found, appears to be dependent upon the particular drug being examined. The complex interrelationship between salt concentration and cooperativity is shown by the observation that decreasing the salt concentration may either increase or decrease the curvature in the initial portion of the binding isotherm. For example, DHAQ shows an increasing degree of cooperativity as the salt concentration is decreased, whereas decreasing the salt concentration decreases the degree of cooperativity for both adriamycin and daunorubicin.

The curvature in the HAAF and NQO binding isotherms (Figure 5) indicates a cooperative binding of these carcinogens to ϕX174 DNA (19). A long range allosteric change in the solution structure of the DNA can explain the observed cooperative binding of these carcinogens. An equally plausible interpretation is that of a cluster model in which two carcinogen molecules bind "together" or cluster into a localized site on the DNA (19).

Krugh and Young (14) have shown shown that daunorubicin facilitates the binding of actinomycin D to poly(dAdT)·poly(dAdT). The following experiments illustrate the interaction of two drugs mediated by a polynucleotide helix. The binding of ^{3}H-actinomycin D to calf thymus DNA at 0.02M NaCl in the presence of fixed concentrations of daunorubicin (r(daunorubicin) = 0.015 or r(daunorubicin) = 0.037) was studied by phase partition. The increasing slope in the binding isotherm of actinomycin D (Figure 3) is diminished when the DNA is pre-equilibrated with daunorubicin, (r(daunorubicin) = 0.015) (Figure 6). At higher concentrations of daunorubicin (r(daunorubicin) = 0.037) the initial increasing slope in the actinomycin D binding isotherm is no longer visible. These two r values correspond to the mid-point of the increasing slope and the peak maximum of the daunorubicin: calf thymus DNA binding isotherm. These results suggest a similar mode of binding for these two antitumor drugs. The kinetic studies

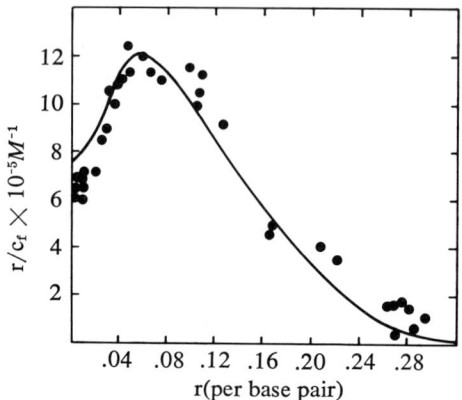

Figure 4. Scatchard plot for the binding of actinomycin D to poly(dGdC) · poly-(dGdC) in 0.01M sodium phosphate, 0.1M sodium chloride, and 0.01M NaEDTA at pH 7.0 and 5°C. Concentrations of the poly(dGdC) · poly(dGdC) (base pairs) ranged from 16μM to 2.3mM, while the actinomycin D concentrations ranged from 0.35μM to 0.64mM.

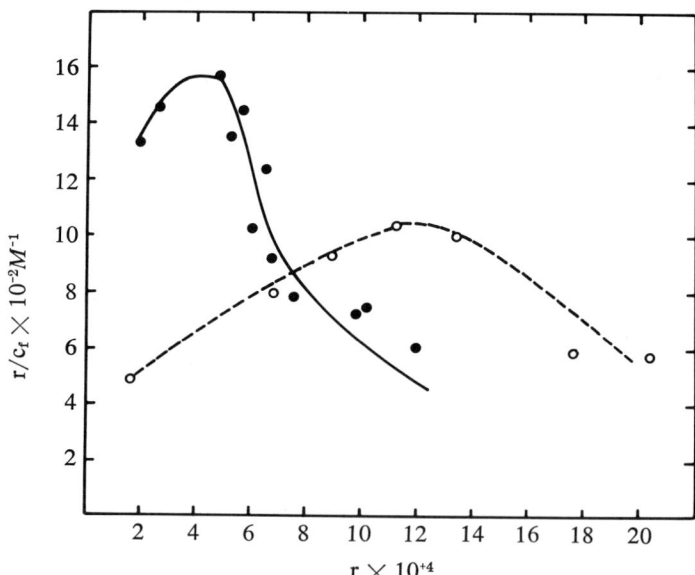

Figure 5. Scatchard plots for carcinogens binding to ϕX174RF DNA. Key: ●-●, HAAF + ϕX174RF; and ○-○, NQO + ϕX174RF; [ϕX174RF] = 0.1–1.0 × 10^{-4}M; $[HAAF]_T = 0.1$–$7.0 × 10^{-5}$M; and $[NQO]_T = 0.05$–$5.0 × 10^{-5}$M. DNA concentrations are in terms of base pairs. The curves through the data are for viewing purposes.

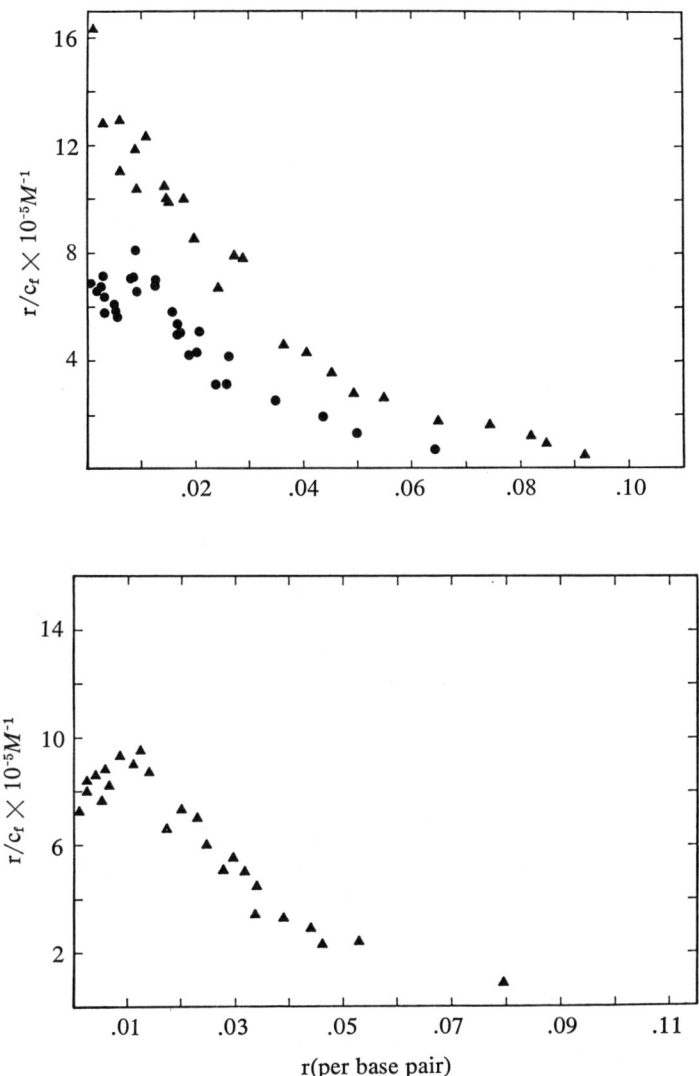

Figure 6. Scatchard plot for the binding of actinomycin D to calf thymus DNA: daunorubicin complex.

Equilibrium binding was measured by the solvent partition method in 10mM sodium phosphate buffer, pH 7.0, 22°C in 0.02M NaCl using ^3H-actinomycin D, partitioning against 1-chloroheptane:n-hexane (1:1). Key ●, DNA daunorubicin ($r = 0.037$) and ▲, DNA:daunorubicin ($r = 0.015$); $[DNA]_T = 0.3$–1.5×10^{-4}M (base pairs); $[daunorubicin]_T = 0.45$–5.4×10^{-6}M; $[actinomycin\ D]_T = 0.02$–$1.0 \times 10^{-5}$M.

on daunorubicin:actinomycin D mixed-drug dissociation from poly(dAdT)·poly(dAdT) also shows that the presence of bound daunorubicin influences the binding of actinomycin D.

Kinetics. The dissociation of actinomycin D from poly(dGdC)· poly(dGdC) is well described by a single exponential decay (Table I). The dissociation of actinomycin D is shown to be dependent on the phosphate to drug ratio (P/D ratio). At high P/D (low degree of saturation) the dissociation lifetime is about 900 sec. At higher degrees of saturation the lifetime increases until a final value of about 1400 sec is reached. Neither the concentration of

Table I
Dissociation lifetime of actinomycin D from Poly(dGdC)·Poly(dGdC) as a function of phosphate/drug ratio (T=22°C) and temperature.

P/D	sec	Temp (°K)	sec
3	1400	290.2	2468
10	1414	295.2	1376
20	1047	300.0	907
30	944	302.1	670
40	914	305.2	529
50	861	310.1	296

SDS nor the pH affects the actinomycin D dissociation lifetime; however, the dissociation time is markedly dependent on both temperature (Table I) and salt concentration (Table II). Decreasing either the salt concentration or the temperature increases the dissociation time, but in neither case does the dissociation change from a single exponential. The divalent metal ion, Mg^{++}, is more effective in lowering the dissociation lifetime than Na^+ (Table II). This suggests that a divalent ion is more effective in lowering the activation barrier than a monocation.

Table II
Dissociation lifetime of actinomycin D from Poly(dGdC)·Poly(dGdC) as a function of Na^+ and Mg^{++} concentration.

$[Na^+]$ (M)	dissociation lifetime (sec)	$[Mg^{++}]$ (M)	dissociation lifetime (sec)
0.053	2624	0.00	2624
0.103	1720	0.02	1098
0.150	1428	0.04	794
0.202	1365	0.08	618
0.322	1261		

The dissociation kinetics of actinomycin D from calf thymus DNA is best fit by at least three exponential decay time constants (Table III), indicating that native DNA has a heterogeneous set of binding sites. The slow dissociation lifetime is associated with the pharmacological activity of this drug and the strongest binding sites. Varying the P/D ratio does not significantly alter this slow dissociation lifetime; however, the percentage of absorbance change associated with this lifetime does depend upon the P/D ratio. At low degrees of saturation the slow dissociation component is the dominant one, accounting for almost 70% of the absorbance change. As the degree of saturation is increased the percentage absorbance associated with this component decreases to about 20% of the total absorbance change. This is due to an increased population of faster dissociating binding sites. Thus the kinetic experiments provide direct evidence for heterogeneity of the binding sites and suggest that the cooperative binding of actinomycin D is associated with the high affinity sites.

Table III.
Relative amplitudes of τ_1, τ_2, and τ_3 for actinomycin D dissociation from calf thymus DNA.

P/D ratio	τ_1 (sec)	τ_2 (sec)	τ_3 (sec)	relative amplitude A (%)	A(%)	A(%)
5	10	70	1600	37.7	43.7	18.6
10	25	90	1410	36.5	38.2	25.4
15	25	80	1292	31.8	41.2	26.9
30	25	81	1250	21.3	40.1	38.5
100	42	216	1527	23.7	15.2	61.1
200	48	305	1517	20.5	12.5	66.6
250	60	486	1600	21.9	10.8	67.3

The dissociation kinetics of daunorubicin and daunorubicin: actinomycin D from poly(dAdT)·poly(dAdT) were studied using stopped-flow techniques. The data is summarized in Table IV. The dissociation of daunorubicin from poly(dAdT)·poly(dAdT) is monophasic with a lifetime of 0.2 sec. In the ternary complex (daunorubicin:actinomycin D:poly(dAdT)·poly(dAdT)) the dissociation of the daunorubicin is characterized by two exponential decays, yielding lifetimes of 0.2 sec and 5 sec. A single 5 sec dissociation time was observed for actinomycin D. This data is consistent with a model in which daunorubicin is bound to isolated DNA binding sites and dissociates with its normal lifetime of 0.2 sec. Whereas daunorubicin and actinomycin D molecules, residing at "adjacent sites", dissociate with a time constant of approximately 5 sec. These results are indicative of a cooperative interaction between daunorubicin and actinomycin D when the two are bound to poly(dAdT)·poly(dAdT).

Table IV
Dissociation of Poly(dAdT)·Poly(dAdT):drug complexes at 22°C.

Poly(dAdT)·Poly(dAdT) :drug complex	Ratio base pairs :drug	τ_1 sec	τ_2 sec
Daunorubicin	20:1	0.2	---
Actinomycin D	20:1	0.7	---
Daunorubicin: Actinomycin D	20:1:1	0.2	5.0

Literature Cited

1. Dattagupta, N.; Hogan, M.; Crothers, D.M. Biochemistry 1980, 19, 5998.

2. Wells, R.D.; Blakesley, R.W.; Burd, J.F.; Chan, H.W.; Dodgson, Hardies, S.C.; Horn, G.T.; Jenson, K.F.; Larson, J.; Nes, I. F.; Selsing, E.; Wartell, R.M. Crit. Rev. Biochem. 1977, 4, 305.

3. Arnott, S.; Chandrasekaran, R.; Birdsall, D.L.; Leslie, A.G. W.; Ratliff, R.L. Nature 1980, 283, 743.

4. Wang, A.H-J.; Quigley, G.J.; Kolpak, F.J.; Crawford, J.L.; van Boom, J.H.; van der Marel, G.; Rich, A. Nature 1979, 282, 680.

5. Patel, D.J.; Canuel, L.L.; Pohl, F.M. Proc. Natl. Acad. Sci. U.S.A. 1979, 76, 2508.

6. Mitra, C.K.; Sarma, M.H.; Sarma, R.H. Biochemistry 1981, 20, 2036.

7. Pohl, F.M.; Jovin, T.M.; Baehr, W.; Holbrook, J.J. Proc. Natl. Acad. Sci. U.S.A. 1972, 69, 3805.

8. Drew, H.; Takano, T.; Tanaka, S.; Itakura, K.; Dickerson R. E. Nature 1980, 286, 567.

9. Pohl, F.M.; Jovin, T.M. J. Mol. Biol. 1972, 67, 375.

10. Hillen, W.; Wells, R.D. Nucleic Acids Res. 1980, 8, 5427.

11. Hogan, M.; Dattagupta, N.; Crothers, D.M. Nature 1979, 278, 521.

12. Müller, W.; Crothers, D.M. J. Mol. Biol. 1968, 35, 251.

13. Krugh, T.R.; Hook, J.W.,III; Balakrishnan, M.S.; Chen, F.M. "Spectroscopic Studies of Actinomycin and Ethidium Complexes with Deoxyribonucleic Acids" Nucleic Acid Geometry and Dynamics: Pergamon Press, New York, 1980, 351.

14. Krugh, T.R.; Young, M.A. Nature 1977, 269, 627.

15. Müller, W.; Crothers, D.M. Eur. J. Biochem. 1975, 54, 267.

16. Albertsson, P.A. Partition of Cell Particles and Macromolecules: Wiley-Interscience, New York, 1971.

17. Waring, M.; Wakelin, L.P.; Lee, J.S. Biochim. Biophys. Acta. 1975, 407, 200.

18. Krugh, T.R.; Winkle, S.A.; Graves, D.E. Biochem. Biophys. Res. Comm. 1981, 98, 317.

19. Winkle, S.A.; Krugh, T.R. Nucleic Acids Res. 1981, 9, 3175.

20. Davanloo, P.; Crothers, D.M. Biochemistry 1976, 15, 4433.

21. LePecq, J.B.; Paoletti, C. J. Mol. Biol. 1967, 27, 87.

22. Wilson, W.D.; Lopp, I.G. Biopolymers 1979, 18, 3025.

23. Jones, R.L.; Lanier, A.C.; Keel, R.A.; Wilson, W.D. Nucleic Acids Res. 1981, 8, 1613.

24. Graves, D.E.; Watkins, C.L.; Yielding, L.W. Biochemistry 1981, 20, 1887.

25. Record, T.M.,Jr.; Anderson, C.F.; Lohman, T.M. Quarterly Rev. Biophys. 1978, 11, 103.

26. Manning, G.S. Quarterly Rev. Biophys. 1978, 11, 178.

RECEIVED October 29, 1981.

INDEX

INDEX

A

Acid, Lewis 15
Acids, hydroxamic 110
ACTH (also see Corticotropin) 120–123
Actinomycin D
   binding to nucleic acids 272
   calf thymus DNA 275, 276f, 279
   dissociation kinetics of
      daunorubicin 272
   DNA, dissociation 270
   poly(dAdt) · poly(dAdt) 275, 279
   poly(dGdC) · poly(dGdC),
      dissociation 270, 278–280
   structure 270, 271f
Acrylamide–HEMA copolymers,
   covalent binding of trypsin
   to cross-linked 134
Acrylate(s)
   fungicidal 37
   8-hydroxyquinolyl 45
   pentachlorophenyl 43
   synthesis
      of 2-(o-benzyl-p-chloro-
         phenoxy)ethyl 49
      of o-benzyl-p-chlorophenyl 48
      of 2-(8-quinolinyloxy)ethyl 48
      of 2-pentachlorophenoxethyl 46
Acryloyloxy-3-4′,5-tribromosalicyl-
   anilide, synthesis 46
2-(2-Acryloyloxethoxy)-3,4′,5-tri-
   bromosalicylanilide, synthesis 47
Acylating agents 137
Adenine, vinyl-type derivatives 196–197
Adipocytes, rabbit 122–123
Adrenocortical cells 122
Adriamycin, binding to nucleic acids 272
Adriamycin structure 270, 271f
Algicides and molluscicides, poly(thio-
   semicarbazide)–copper(II)
   complexes as potential 55–72
Allosterism in binding of various
   antibiotics and carcinogens
   to DNA 269–280
Amine oxides, vinyl 3
Aminoacridine dyes bound to DNA .. 177
p-Aminobenzenesulfonyl carbamates
   of hydrogels (p-ABSC-
   HEMA) 133–147
Amphibian melanophores 120, 122–123

Anemia, Cooley's, design of polymeric
   iron chelators for treating
   iron overload 107–117
9,10-Anthracenedione,
   substituted 270, 271f
Antibiotics, anthracycline, binding
   isotherms for interaction with
   calf thymus DNA 273, 274f
Antibiotics and carcinogens, cooperativity and allosterism in binding
   to DNA 269–280
Anticancer agents, complexes of
   cis-diamminedichloroplatinum(II)
   and α-hydroxyquinones, con-
   trolled release 233–240
Antigens 214
Antimitotic agents 78
Antimony-containing polymers 17
Antineoplastic agents 3
Antineoplastic agents, polymeric
   derivatives based on cis-di-
   chlorodiammineplatinum II 221–230
Antineoplastic polymers 195–201
Antitumor
   action of pyran copolymer 164
   activity, polymers 195
   activity, related to molecular
      weight, macrophage activation 208t
   effects of polycarboxylic acid
      polymers 205–218
   drugs 270, 271f
Antiviral encephalomyocarditis
   protection with polymer
   molecular weight, change 207t
Arsenic-containing polymers 17
Arylsulfonyl carbamates 136, 137
Arylsulfonyl isocyanates 137
Assay(s)
   methods, enzyme kinetic 144
   paper disk 14
   protein 14
   slant culture 14
Aureobasidium pullulans 36

B

Bacteria, effect of uranyl ion 24t
Bacteria inhibition as function of
   organostannane polyxanthene 21t

Bacterial endotoxin, effect of molecular weight of pyran on acute toxicity and sensitivity 213t
Base, Lewis 15
2-Benzyl-4-chlorophenol 37
o-Benzyl-p-chlorophenyl acrylate, synthesis 48
2-(o-Benzyl-p-chlorophenoxy)ethanol, synthesis 48
2-(o-Benzyl-p-chlorophenoxy)ethyl acrylate, synthesis 49
2-(o-Benzyl-p-chlorophenoxy)ethyl bromide, synthesis 51
Bioactive polymers 1–8
Bioagents, polymeric 75–81
Biocidal activity of organotin polymers in wood 27–33
Biocides, polymer-bound 36
Biological analysis techniques 14
Biologically active macromolecules, advantages 7
Blood cell membrane, interaction of pyran copolymer with human red 164
Bovine serum albumin (BSA) 99
from hemisphere systems, release kinetics 101, 102f
methotrexate covalently linked 255
2-(2-Bromoethoxy)-3,4′,5-tribromosalicylanilide, synthesis 50

C

Ca^{2+} 169, 170f, 171
Cancer chemotherapy 205
Cancer therapy, polymers 193–201
Carbamate(s)
  activity analysis for trypsin CVB hydrogel sulfonyl 144t
  arysulfonyl 137
  herbicides 78
  hydrogel 133–147
  insecticide 80
Carbohydrate groups of the glycoprotein hormones 129
Carcinogens 269–280
Carcinoma, polymer activity against Lewis lung 206
Catechols 110
Cell membranes, interaction of synthetic polymers 163–174
Cellulose compounds, organostannane-containing, growth of organisms 23t
Cellulose-containing tissues 150
Chelation therapy, iron drugs 111
Chelators, iron
  bioassay of polymeric 114, 115t
  design 110–111

Chelators, iron (continued)
  formation constants 108t
  potential advantage of polymeric 114, 116
  treating iron overload in Cooley's anemia, design of polymeric 107–117
  naturally occurring 108, 109f
Chemotherapeutic synthetic polymers 2
Chemotherapy, cancer 205
Chemotherapy drugs, design 4, 14
α-Chitin 150, 151f
β-Chitin 150
Chitin
  core-sheath model 156, 157f
  polysaccharide 149
  –protein complexes 149–160
4-Chlorophenyl isocyanates 78
Chorionic somatomammotropin (hCS), human 128
Collagen 149, 150
Controlled-release organotin antifouling polymers 27–33
Cooley's anemia, design of polymeric iron chelators for treating iron overload 107–117
Copper speciation 55–72
Corticotropin (ACTH) 119
Corticotropin and melanotropins, structure 121f
Cotton 21
Creosote 29
Cuticles, insect 150
Cyanogen bromide technique for activating hydroxyl-containing polymeric supports 134
Cyclophosphamide, derivatives 3
Cytostasis by pyran-activated macrophages 214
Cytotoxicity of activated macrophage 215

D

Daunorubicin
  binding of actinomycin D to poly(dAdT) · poly(dAdT), effect 275
  binding isotherms for the interaction with calf thymus DNA 273, 274f
  binding to nucleic acids 272
  dissociation kinetics 272, 279
  structure 270, 271f
Decay, microbiological, protection of wood against 27–33
Decay, wood 32
Desferrioxamine-B 108
Dextran 21

INDEX

DHAQ 270, 271f
DHAQ, binding isotherm for the
  interaction with chalf thymus
  DNA 273, 274f
cis-Diamminedichloroplatinum(II)
  as antineoplastic agent, polymeric
    derivatives 221–230
  controlled release of anticancer
    agents 233–240
  IR spectrum 225f
  polymeric derivatives 200–201
  synthesis and physical characteri-
    zation 223–224
2,3-Dicarboxynorborn-5-ene copoly-
  mers against Ehrlich ascites
  tumor, effect 210t
Dihydrofolate reductase, effect of
  MTX(poly-D-lysine) and
  MTX(poly-L-lysine) on rat
  hepatocyte 264, 266t
Dihydrofolate reductase and thy-
  midylate synthetase, activities in
  MTX and MTX(poly-L-lysine)
  treated H35 cells, measure-
  ment 260, 261f
1,4-Dihydroxy-9,10-anthraquinone
  (DHAQ), synthesis and characterization of complexes 234
5,8-Dihydoxy-1,4-naphthoquinone
  (DHNO), synthesis and charac-
  terization of complexes 234
Dimethylolurea, copolymers of sulfa
  drugs 2
2,3-Dimethylsuccinic acid 246, 248
Dimethylsuccinic acid and anhydride,
  ^{13}C NMR data 249f
2,5-Dimethyltetrahydrofuran-3,4-di-
  carboxylic acid 246, 247f, 251
Dipalmitoylphosphatidylcholine
  (DPPC) 171
Dipalmitoylphosphatidylcholine
  (DPPC), interactions of pyran
  copolymer with liposomes 164
Disk, paper, assays 14
Distamycin, binding of netropsin
  to calf thymus DUA 270
Divinyl ether–maleic anhydride
  copolymer (DIVEMA) 3, 163, 196,
                                            243–253
DNA
  aminoacridine dyes bound 177
  analogs 3
  base composition, interaction
    heat effect 187
  binding
    actinomycin to calf
      thymus 275, 276f
    antibiotics and carcinogens,
      cooperativity and allo-
      sterism 269–280

DNA (continued)
  binding (continued)
    dye, effect of GC base
      composition 177–189
    ethidium to calf thymus 270
    isotherm for interaction of
      DHQA with calf
      thymus 273, 274f
    netropsin and distamycin to
      calf thymus 270
    dissociation of actinomycin D 270
    dissociation kinetics of actino-
      mycin D from calf thymus 279
    –dye complex thermodynamic
      quantities 179–187
    effects of salt concentration on
      binding of drugs 275
    and PF, heats of mixing 179
    thermodynamic characterization of
      proflavin binding 177–189
DOXO, synthesis and characterization
  of complexes 234
Doxorubicin base from doxorubicin
  hydrochloride preparation 236
Drug(s)
  antitumor 270, 271f
  carrier-bound 255
  chemotherapy, design 4, 14
  delivery systems for macro-
    molecules, polymeric 95–103
  delivery systems, mem-
    brane-enclosed 90
  with dimethylolurea, copolymers
    of sulfa 2
  to DNA, effects of salt concentra-
    tion on binding 275
  with formaldehyde, copolymers of
    sulfa 2
  polymeric 4–8, 194
  –polymer-sustained release
    systems 85–93
  release, pharmacokinetic models
    of sustained 86–90
  sulfa 200
  therapy, polymeric 193–195
  use in iron chelation therapy 111
Dye(s)
  binding by intercalation
    process 177–189
  bound to DNA, aminoacridine 177
  complex, thermodynamic quantities
    of DNA 179–187
  xanthene, tin-containing 17

E

EDTA 64
Ehrlich ascites tumor cell(s)
  effect of 2,3-dicarboxynorborn-
    5-ene, copolymers 210t

Ehrlich ascities tumor cell(s) (*continued*)
  effect of maleic anhydride
    copolymers 209*t*
  proliferation in vivo by pyran
    inoculation, inhibition 217*t*
Encephalomyocarditis
  effect of platinum polyamines 228–230
  polymer activity against 206
  protection with polymer molecular
    weight, change in antiviral 207*t*
Endocarditis, prosthetic valve 86
Endorphin (EP) 119, 122–125
Endotoxin, bacterial, effect of
  molecular weight of pyran on
  acute toxicity and sensitivity 213*t*
Endotoxin sensitization with polymer
  molecular weight, effect 211*t*
Enkephalin, methionine 123
Enterobactin 108, 109*f*
Enzyme
  apparent Michaelis constant of
    immobilized 146
  conjugate, hydrogel 147
  kinetic assay methods 144
EP (*see* Endorphin)
Erythrocyte(s)
  binding of pyran copolymer
    to intact 165
  effect of Ca²⁺ on binding of pyran
    copolymers to intact 170*f*
  ghosts, effect of divalent ions on
    interaction of pyran 169
  ghosts, effect of pyran on heat
    capacity, profiles of
    human 166, 167*f*, 168*f*
Escherichia coli 21
Ether–maleic anhydride cyclic alternating copolymers, ¹³C NMR
  studies 243–253
Ethidium to calf thymus DNA,
  binding 270
Ethylenediamine-*N,N'*-bis(2-
  hydroxyphenyllactic acid) 111, 112*f*
Ethylene–maleic anhydride,
  copolymer 195
Ethylene–vinyl acetate copolymer
  (EVA) 95–103

F

Ferrichrome 108, 109*f*
Fiber–matrix composites 149
5-Fluorouracil 3, 198, 199
Folinic acid 257, 267
Follitropin (FSH) 119, 128
Formaldehyde, copolymers of sulfa
  drugs 2
FSH (*see* Follitropin)
Fungicidal acrylates 37

Fungicidal monomer synthesis 37–40
Fungicides for paints,
  polymer-bound 35–53
Fungi-related rot and mildew,
  retardation 21, 23

G

Gentamicin from prostetic heart
  valves, sustained release 86–90
Glutaraldehyde immobilization of
  trypsin 136–137
Glycoprotein hormones 128–129
Gonadotropins 128
Growth hormone (GH) 126–128

H

³H-Inulin from ethylene–vinyl
  acetate copolymer 98*f*
HAAF 271*f*, 272
Halides, organostannane, with poly-
  saccharides, condensation 21
Heart valves, sustained release of
  gentamicin from prosthetic 86–90
HEMA copolymers covalent binding
  of trypsin to cross-linked
  acrylamide 134
HEMA/MMA copolymers, sustained
  release of tetracycline 90–92
Hemisphere polymer systems 97–101
Hemisphere-shaped devices for
  release of macromolecules 100*f*
Hemoglobin, β-chain 107
Hepatocyte(s)
  dihydrofolate reductase, effect of
    MTX(poly-D-lysine) and
    MTX(poly-L-lysine) 264, 266*t*
  and hepatoma cells to
    MTX(poly-L-lysine) 267
  toxicity of poly-L-lysine and
    poly-D-lysine derivatives 264, 265*f*
  uptake of MTX and
    MTX(poly-L-lysine) 260
Hepatoma cell(s) 260
  growth and dihydrofolate reductase,
    effect of MTX(poly-L-lysine)
    pulse doses 264, 266*t*
  interaction of MTX–BSA with
    MTX transport-resistant
    Reuber H35 255
  to MTX(poly-L-lysine), reponse
    of hepatocytes 267
Herbicides
  carbamate 78
  controlled release 80
  derived from poly(vinyl alcohol),
    polymeric 75–81
Hexobarbitol sleeping time effected
  by polymer molecular weight 212*t*

INDEX

Hormone(s)
  glycoprotein 128–129
  growth, (GH) 126–128
  lactogenic 128
  peptide 120
  pituitary, structural and functional interrelationships of anterior 119–129
  placental 128
  protein 125
Hydrogel(s) carbamates 133–147
  -containing polymeric supports, cyanogen bromide technique for activating 134
  –enzyme conjugate 147
  mechanical properties 137
  modification 138t
  p-nitrobenzenesulfonyl carbamates, reduction 136
  p-nitrophenyl carbamates, preparation 134
  sulfonyl carbamates, activity analysis for trypsin CVB 144t
  synthesis 135t
  system, styrene content 147
Hydroxyamic acid 110
  complexes, iron 116
  polymers 111
5-Hydroxyl-1,4-naphthoquinone (HNO), synthesis and characterization of complexes 234
8-Hydroxyquinoline 37
8-Hydroxyquinolyl acrylate 45
α-Hydroxyquinones, controlled release of anticancer agents 233–240
Hydroxyquinone ligands and their Pt(II) complexes, IR absorption, UV spectra 237t

I

Immunopotentiators, macrophage activation by polycarboxyl acid polymer 214–217
Insect chitin–protein complex from ovipositor of ichneumon fly *Megarhyssa* 149–160
Insect cuticles 150
Insecticide carbamates 80
Intercalation process, dye binding 177–189
Interferon 205
Inulin from ethylene–vinyl acetate copolymer matrices 95–103
Ionophore, macromolecular 164
Ions, effect of divalent, on interaction of pyran with erythrocyte ghosts 169
Iron chelating ability of desferrioxamine-B 108

Iron chelators
  bioassay of polymeric 114, 115t
  design 110–111
  formation constants 108t
  naturally occurring 108, 109f
  potential advantage of polymeric 114, 116
  therapy, drugs used 111
Iron–hydroxamic acid complexes 116
Iron overload in Cooley's anemia, design of polymeric iron chelators for treating 107–117
Isocyanates 75–81, 137

K

Kinetic(s)
  assay methods, enzyme 144
  BSA from hemisphere system, release 101, 102f
  inulin from ethylene–vinyl acetate copolymer matrices, release 95–103
  release, comparison of in vitro and in vivo 96–97
  release, zero-order 97–101

L

Lactices, synthesis of stable 52–53
Lactogenic hormones 128
Latices, terpolymer 40
Lewis acid, base 15
Lewis lung carcinoma, polymer activity against 206
LH (*see* Lutropin)
Lipid surfaces, interaction of pyran copolymer with membrane 163–174
Liposomes 171
Lipotropin (LPH) 119, 123–125
LPH (*see* Lipotropin)
Lung carcinoma, polymer activity against Lewis 206
Lutropin (LH) 119, 128

M

Macromolecular ionophore 164
Macromolecules advantages of biologically active 7
Macromolecules, polymeric drug delivery systems 95–103
Macrophage(s)
  activation 215
  antitumor activity and antiviral activity, relationship between molecular weight 208t
  by polycarboxyl acid polymer immunopotentiators 214–217
  cell surface antigen 215

Macrophage(s) (continued)
  cytostasis by pyran-activated 214
  cytotoxicity of activated 215
  pyran-activated 164
  tumoricidal activity, mechanism
    of activated 216
Malachite 64
Maleic anhydride copolymer(s)
  against Ehrlich ascites tumor
    cells, effect 209t
  5- and 6-membered cyclic struc-
    tures in divinyl ether 244
  ^{13}C NMR spectrum of hydrolyzed
    divinyl ether 245f
Maleic anhydride (pyran copolymer),
  1:2 copolymer of divinyl ether .. 163
Matrix material, polymer 6
Medicines, metal-containing
  polymers 14
Megarhyssa lunator
  model for three-dimensional
    structure of typical insect
    chitin–protein complex from
    ovipositor fo the ichneumon
    fly 149–160
  ovipositor, structure of
    chitin 150, 151f
  x-ray fiber diagram of intact ovi-
    positor 153, 154f
Melanin dispersing agent 120, 122–123
Melanin in amphibian
  melanophores, dispersion 120
Melanophores, amphibian 120, 122–123
Melanotropins 119, 120, 121f
Membranes, interaction of synthetic
  polymers with cell 163–174
6-Mercaptopurine 3, 198, 199
Mercurochrome 21
Metal-containing polymers,
  biological activities 3, 13–24
Methionine enkephalin 123
Methotrexate (polylysine) with rat
  liver tumor cells interaction ... 225–267
Methylbenzene, substituted 111, 112f
Michaelis constant, of immobilized
  enzymes 146
Microbiological decay, protection of
  wood against 27–33
Mildewcidal coating polymers 35, 36
Molecular weight, macrophage acti-
  vation, antitumor activity and
  antiviral activity, relationship 208t
Molluscicides, poly(thiosemicarba-
  zide)copper(II) complexes as
  potential algicides 55–72
MSH 120–123
α- and β-MSH (see Melanotropins)
MTX (see Methotrexate)

N

Netropsin and distamycin to calf
  thymus DNA, binding 270
Niridazole from silicone rubber,
  sustained release 93
Nitrate, uranyl 24
p-Nitrobenzenesulfonyl carbamates,
  reduction of hydrogel 136
p-Nitrobenzenesulfonyl
  isocyanate 134, 137
p-Nitrophenyl carbamates,
  preparation of hydrogel 134
p-Nitrophenyl-p'-guanidino-
  benzoate (NPGB) as active-site
  titrant for trypsin 141
NQO 271f, 272
Nucleic acid analogs as antineo-
  plastic polymers 196–199
Nucleic acids, binding 272

O

Opiate peptides, endogenous 123
Organotin antifouling polymers,
  controlled-release 27–33
Organotin, monomers, in situ
  polymerization 29, 30t
Organotin polymers in wood,
  biocidal activity 27–33
Organostannane
  -containing cellulose compounds,
    growth of organisms 23t
  dihalides, condensation with
    xanthane salts 17
  halides, condensation with
    polysaccharides 21
  -modified polysaccharide com-
    pounds, inhibition 23t
  polyxanthene, bacterial inhibition
    as function 21t
Ovipositor of Megarhyssa lunator,
  x-ray fiber diagram 153, 154f
Oxides, vinyl amine 3

P

Paint(s)
  defacing organism 36
  films against mildew, protection 35
  polymer-bound fungicides 35–53
Pentachlorophenol 36
Pentachlorophenyl acrylate 43
2-Pentachlorophenoxethyl acrylate,
  synthesis 46
2-Pentachlorophenoxyethyl bromide,
  synthesis 49
2-Pentachlorophenoxyethanol,
  synthesis 45–46
Peptide hormones 120

INDEX

Peptides, endogenous opiate 123
PF, heats of DNA mixing 179
Pharmacokinetic models of sustained
  drug release 86–90
Phase partition analysis 272
PHEMA [see Poly(hydroxyethyl methacrylate)]
Phenols 110
Phospholipids, interactions of pyran
  copolymer 163–174
Phosphorus-containing polymers 3
Pituitary hormones, structural and
  functional interrelationships
  of anterior 119–129
Placental hormone 128
Plasmin 126
Platinum compounds, toxicity 227t, 228t
Platinum polyamines 221–230
Poliovirus Type I, effect of
  platinum polyamines 228–230
Poly(acrylic acid) 195, 206
Polyamines, platinum 221–230
Polyanionic antineoplastic
  polymers 195–196
Polyanionic systems, mechanism of
  tumor inhibition 196
Polyanions 205
Polycarboxylic acid polymers
  antitumor and antiviral effects 205–218
  clinical effects 218
  immunopotentiators, macrophage
    activation 214–217
  toxicological effects 210–213
Poly[cis-dichloro(1,6-hexanedi-
  ammine)platinum(II)], IR
  spectrum 225f
Poly(dAdT) · poly(dAdT) 275, 279, 280t
Poly(dGdC) · poly(dGdC), binding of
  actinomycin D 275, 276f
Poly(dGdC) · poly(dGdC), dissocia-
  tion of actinomycin D 270, 278–280
Polydyes, tin 17
Polyesters, uranyl 24
Polyether esters, titanium-
  containing 15, 17
Poly(α-ethylacrylic acid) 166
Poly(2-hydroxyethyl methacrylate),
  hydrolysis 139f
Poly(hydroxyethyl methacrylate),
  (PHEMA) 133
Poly(hydroxypropyl acrylate),
  hydrolysis 139f
Poly(lactic acid) (PLA) 234
Polylysine, interaction of metho-
  trexate with rat liver tumor
  cells 255–267
Poly(maleic acid) (PMA) 206
Polymer(s)
  antimony-containing 17
  antineoplastic 199–201

Polymer(s) (continued)
  with antitumor activity 195
  antitumor and antiviral effects
    of polycarboxylic acid 205–218
  arsenic-containing 17
  bioactive 1–8
  biological activities of
    metal-containing 13–24
  bound biocides 36
  bound fungicides for paints 35–53
  for cancer therapy 193–201
  clinical effects of polycarboxylic
    acid 218
  condensation, metal-containing 14
  (DIVEMA), pyran 3
  hydroxamic acid 111
  implant 97
  matrix material 6
  medicines, metal-containing 14
  metal-containing 3
  mildewcidal coating 36
  molecular weight
    change in antiviral encephalo
      myocarditis protection 207t
    effect on acute toxicity 211t
    hexobarbitol sleeping time
      effected by 212t
    inhibition of tumor size 206t
  nucleic acid analogs as
    antineoplastic 196–199
  organotin, in wood biocidal
    activity 27–33
  phosphorus-containing 3
  polyanionic antineoplastic 195–196
  sustained release systems 85–93
  synthetic
    bioactive 2
    chemotherapeutic 2
  interaction with cell
    membranes 163–174
  systems, hemisphere 97–101
  tin-containing 17
  toxicological effects of poly-
    carboxylic acid 210–213
  trialkyltin 27–33
Polymeric
  bioagents 75–81
  drug(s) 4–8
  delivery systems for macro-
    molecules 95–103
  soluble, insoluble 194
  therapy 193–195
  iron chelators for treating iron
    overload in Cooley's anemia,
    design 107–117
  matrices, release of macro-
    molecules 95–103
Polymerization of organotin
  monomers, in situ 29, 30t
Poly(methacrylic acid) 195

Poly(N-methacryloyl-β-alanine-
  hydroxamic acid) 111, 112f
Poly(N-methacryloylhydroxamic
  acid) 113
Polyoximes, uranyl 24
Polysaccharide chitin 149
Polysaccharide sources 21–23
Poly(thiosemicarbazide)copper(II)
  complexes as potential algicides
  and molluscicides 55–72
Poly(thiosemicarbazides),
  preparation 56
Poly(vinyl alcohol), polymeric
  herbicides derived 75–81
Polyxanthene, organostannane,
  bacteria inhibition 21t
Proflavin binding to DNA, thermo-
  dynamic characterization 177–189
Prolactin 128
Proopiomelanocortin 125
Prosthetic heart valves, sustained
  release of gentamicin 86–90
Protein assays 14
Protein complexes, chitin 149–160
Protein hormones 125
Pseudomonas aeruginosa 21
cis-Pta₂Cl₂-doxorubicin
  (DOXO) 234, 236, 238, 239f
cis-Pta₂Cl₂/5-hydroxy-1,4-naptho-
  quinone, preparation 235
Pt(II) complexes, UV spectra of
  hydroxyquinone ligands and
  IR absorption 237t
Pyrimidine compounds, vinyl
  derivatives 2
Pyran
  activated macrophages, cytostasis .. 241
  acute toxicity 211, 213t
  copolymer 163–174, 205
  inoculation, inhibition of Ehrlich
    ascites tumor cell prolifera-
    tion in vivo 217t
  isomer 253
  polymer (DIVEMA) 3

Q

2-(8-Quinolinyloxy)ethanol,
  synthesis 47–48
2-(8-Quinolinyloxy)ethyl acrylate,
  synthesis 48

R

Rabbit adipocytes 122–123
Red blood cell membrane, interactions
  with pyran copolymers 164
Release of macromolecules from
  polymeric matrices 95–103

Rhodotorulic acid 108, 109f
RNA, analogs 3
RNA viruses 228–230
Rubber, silicone, sustained release
  of niridazole 93

S

Salt concentration on binding of
  drugs to DNA, effects 275
Sarcolysin 199
Schistosomiasis 55, 93
Siderochromes 108, 109f
Silicone rubber, sustained release
  of niridazole 93
Slant culture assays 14
Sodium poly(vinyl sulfonate) 195
Somatomammotropin, human
  chorionic 128
Styrene content in hydrogel system 147
Sulfa drugs 2, 200
Sulfonyl carbamates, activity analysis
  for trypsin CVB hydrogel 144t
Sustained release systems, in vivo
  studies on drug–polymer 85–93
Synthon, p-nitrobenzenesulfonyl
  isocyanate 134

T

Tetracycline from HEMA/MMA
  copolymers, sustained release 90–92
Tetrahydrofuran diethylester 251
Tetrahydropyran 246
Tetra-S-carbamidomethylated hGH .. 126
Thymidylate synthetase activities in
  MTX and MTX(poly-L-lysine)
  treated H35 cells, measurement
  of dihydrofolate reductase 260, 261f
Thyrotropin (TSH) 119, 128
Tin-containing polymers 17
Tin, polysaccharides, containing 21–23
Tissues, cellulose-containing 150
Titanium-containing polyether
  esters 15, 17
Toxicological effects of poly-
  carboxylic acid polymers 210–213
Trialkyltin polymers 27–33
Triazacyclotridecane,
  substituted 111, 112f
Triazine compounds, vinyl
  derivatives 2
3,4′,5-Tribromosalicylanilide 37
Tributyltin 28, 29
Trypsin
  covalent binding to hydrogels 133–147
  immobilization 140–141
  p-nitrophenyl-p′-guanidino-
    benzoate as active-site titrant .. 141
TSH (see Thyrotropin)

Tumor(s)
cell(s)
effect of anhydride copolymers against Ehrlich ascites 209t
interaction of methotrexate (poly-lysine) with rat liver 255–267
proliferation in vivo by pyran inoculation, inhibition of Ehrlich ascites 217t
effect of 2,3-dicarboxynorborn-5-ene copolymers against Ehrlich ascities 210t
inhibition by polyanionic systems, mechanisms 196
size with polymer molecular weight, inhibition 206t
Tumoricidal activity, mechanism of activated macrophage 216

U

Uracil, vinyl-type derivatives 196–197
Uranyl-containing compounds 23–24

V

Vinyl
amine oxides 3
derivatives of pyrimidine and triazine compounds 2
ethers, various, synthesis 49–52
2-Vinylpyridine-1-oxide 4
N-Vinylpyrrolidone 3
Viruses, effect of platinum poly-amines on RNA 228–230

W

Water, composition of synthetic natural 58
Wood, biocidal activity of organotin polymers 27–33

X

Xanthane salts, condensation of organostannane dihalides 17
Xanthene dyes, tin-containing 17

Jacket design by Kathleen Schaner.
Editing and production by Susan Moses and Karen Gray.

Elements typeset by Service Composition Co., Baltimore, MD.
Printed and bound by Maple Press Co., York, PA.